Eric Kaplan is an executive producer and writer for *The Big Bang Theory*. Previously he wrote for *The Simpsons, Futurama* and *Flight of the Conchords*. He graduated from Harvard and is currently completing his dissertation in philosophy at the University of California, Berkeley.

www.ericlinuskaplan.wordpress.com

'One of the most enjoyable and thought-provoking books I've ever read. Eric Kaplan will convince you that comedy provides as much insight as logic or mysticism into the fundamental nature of reality' Sean Carroll, theoretical physicist at Caltech and author of *The Particle at the End of the Universe*

'Kaplan's deadpan style lets you read this as a serious philosophical treatise – but you can also take it as a well-done farce' *New York Post*

'If you can put this book down, you should see a doctor. Kaplan's message burrows into the mind, beats up a few beliefs and then leaves with a triumphant bang' Michael Gazzaniga, Professor of Psychology University of California Santa Barbara, Director of the SAGE Center for the Study of the Mind, and Founder of the Cognitive Neuroscience Society

'Exceptionally interesting, rigorous and I found it not only weirdly funny but deeply moving' Hubert Dreyfus, Professor of Philosophy, University of California Berkeley, Fellow of the American Academy of Arts and Sciences

DOES SANTA EXIST?

A Philosophical Investigation

Eric Kaplan

ABACUS

ABACUS

First published in the United States in 2014 by Dutton
First published in Great Britain in 2014 by Little, Brown
This paperback edition published in 2015 by Abacus

1 3 5 7 9 10 8 6 4 2

A CIP catalogue record for this book is available from the British Library.

ISBN 978-0-349-14062-9

Printed and bound in Great Britain by
Clays Ltd, St Ives plc

Papers used by Little, Brown are from well-managed forests and other responsible sources.

MIX
Paper from responsible sources
FSC® C104740

Little, Brown
An imprint of
Little, Brown Book Group
Carmelite House
50 Victoria Embankment
London EC4Y 0DZ

An Hachette UK Company
www.hachette.co.uk

www.littlebrown.co.uk

For Răduca

CONTENTS

A NOTE TO THE READER

This book has three titles designed by the writer to appeal to three different audiences, but as luck would have it, all three look the same in contemporary written English, viz.: Does Santa Exist?

The first title, *Does Santa Exist?*, with the words stressed more or less equally, is for readers who would like to consider whether or not the kindly, bearded gift giver exists.

The second title, *Does Santa Exist?*, should be read with a rising stress on the final word, sort of like this: "Does Santa *Exist*?" This title is a lure for readers who would like to know whether, if Santa does do anything in the ontological line, that thing he does qualifies as existing. Their interest, in other words, lies less in Santa and more in the concept of existence.

The third title, *Does Santa Exist?*, should be read as the

conclusion of the following very short little dialogue on the relationship between science and faith:

A: Just because something is not part of scientific discourse doesn't mean it can't still exist.

B: Really? Does *Santa* Exist?!

For readers drawn to this title, the question "Does Santa Exist?" should be pronounced as if it is a reductio ad absurdum argument against the claim that things exist that are beyond the ken of science. Their chief interest in this book, presuming they have any, is as a consideration of whether or not that reductio argument is valid.

ELK

INTRODUCTION

My Son, His Friend's Mother, and Two Explanations

The ontology of Santa Claus didn't impinge on my life until my son, Ari, was in kindergarten. Ari did not believe in Santa Claus. He was supposed to go to the zoo in early December with his friend Schuyler, and Schuyler's mother, Tammi, called me up and said she didn't want her son to go because there

were reindeer there, and reindeer, she felt, would lead to a discussion of Santa Claus. Tammi's son, Schuyler, did believe in Santa Claus: He was still firmly a sweet child and not yet in sour and rebellious teenager territory, and she wanted him, at least for a while, to stay that way. So Tammi wanted to cancel the playdate to ensure that Ari would not tell her son, "There is no Santa—he's just your parents," and shake his belief.

I found this a troubling interaction because I thought Tammi was sacrificing her son's friendship with Ari, who was real, in order to preserve his relationship with Santa Claus, who was not.

Why was I so sure he didn't exist? Not because I've never seen him—I've never seen Israeli supermodel Bar Refaeli and she exists, or at least did as of this writing. And not because if I went to the North Pole, I wouldn't see him and his elves—just a lot of snow and ice and so forth—because there are any number of explanations that would square with that. Santa might emit a field from his beard that makes people miss him, the elves might have a machine that causes light to bend, or I could have met him and then been convinced by Mrs. Claus to undergo brain surgery that erased my memory. No, the real reason, I'm sure, is that nobody had ever told me he did, and belief in Santa Claus did not fit in with a number of other things I knew to be true—e.g., reindeer don't fly, toys come from the store, etc.

I told this story to my daughter and she said, "I believe in Santa Claus." I also asked her if she believed in the Easter bunny and she said, "Yes. I'm a kid, so I believe in everything."

I told this story to my wife, who is a psychologist raised in Communist Romania, and she said something along the lines of "American parents lie to their children about this stupidity, and then the children grow up and find out their parents lied to them. No wonder American children are screwed up."

I remained puzzled by Tammi's behavior. I could think of two possible solutions:

THE LIAR EXPLANATION

For some reason back in the past, American children were taught to believe in Santa Claus—probably because their parents thought it was a good way to scare them into being good. When the children grew up and stopped believing in Santa Claus, they decided it would be a good idea to trick their children into believing. So society is basically divided into two groups of people—the liars and the lied to. The liars have motivations ranging from the benevolent (parents presumably) to the self-interested (the sellers of Christmas merchandise, American politicians who want a national myth that will unite a nation of immigrants). Let's be blunt and call this the LIAR story.

I've observed evidence that the LIAR story is true. I work in Hollywood, which pumps a lot of images and stories out into the consciousness of the globe. When we were writing an episode of a television show called *The Big Bang Theory,* in which the character Sheldon kills Santa Claus in a *Dungeons and Dragons* game, one of the writers wanted to be sure that our story left the existence of Santa Claus open, because his kids were going to watch the show and they believed in Santa Claus. Of course, since he was a writer for a U.S. sitcom that is supported by commercials, his benevolent motivations for lying meshed with the less benevolent motivators of our advertisers.

THE CRAZY EXPLANATION

Another solution to the puzzle was that something in Tammi's mind is divided or dissociated. So, according to this theory, it's possible that a part of Tammi's mind does believe in Santa Claus. She doesn't talk about it when she talks to other adults, but when alone with her child, she believes. The part of Tammi that believes in Santa might not even be a part that has access to her mouth. So she might never say, "I believe in Santa Claus," but she is disposed to have dreams, fantasies, and feelings related to Saint Nick. As a consequence, she is uncomfortable with having her son lose faith in Santa Claus because some system in her brain believes too.

How can one person believe and not believe in Santa Claus? If you are a strong proponent of the conspiracy story, you may not believe this is the case—you might think that if she ever does confess to Santa belief, she is just lying. After all, she buys toys at the store—how can she honestly maintain they come down the chimney?

But people believe different things at different times and in different contexts. Let's imagine Tammi goes home and goes to bed. As she drifts off to sleep, she hears a voice in her head, one that sounds like her own. It says, "Santa does exist. I remember waiting for him to come. How do I know he didn't? Yes, part of me thinks he didn't come and never will, but why should I listen to that part?"

Tammi has a couple of different Tammis inside her. She has the Tammi who once believed in Santa but now buys toys from the store, and she has the Tammi who still does believe in Santa. This Tammi feels good when she thinks about Santa and angry when she thinks about Eric not believing in Santa. This Tammi can effortlessly respond to Santa images and Santa television shows and songs about Santa.

Tammi's self could be divided; she could be more than one

of her Tammis at the same time—that is, she could have one voice in her head that says, "Of course Santa Claus does not exist," and another voice that says, "I hope he brings me something good!" Or her self could be divided *across time.* That is, she could make fun of Santa Claus all year long until Christmas season and then talk during Christmas as if she does believe in the jolly old saint.

Since it invokes voices in the head, let us call this, uncharitably, the CRAZY explanation.

The LIAR and the CRAZY explanations are similar on a deep level because while LIAR appeals to dissociation on the interpersonal level, CRAZY appeals to dissociation on the intrapersonal level. Societies run by conspiracies built on lies are schizophrenic; crazy people lie to themselves.

In the CRAZY explanation, there is some kind of disunity within Tammi—there is a part of her that believes and a part that doesn't believe. In the LIAR explanation, there is a disunity in America—there is part that believes and part that doesn't believe. And in both, there is something sort of screwed up about the relationship among these parts. You can even switch the explanations. You can say that Tammi is lying to herself, or that America is a little crazy on the subject of Santa Claus.

Is the LIAR or the CRAZY explanation correct?

Versions of both of them are found throughout rationalist critiques of religion and scientific accounts of human behavior in general. For example:

- Marxism—Liar. Priests lie to people to keep the powerful in power: "There'll be pie in the sky when you die."
- Psychoanalysis—Crazy. People's minds create irrational beliefs to defend against all the psychic pressure they're under, what with death and wanting to sleep with their mothers and so on.
- Neurobiology—Crazy. People have evolved modules in their brains that perceive humans as existing because it was evolutionarily important to know if somebody was in your cave with you. When we think Santa exists, it's because that chunk of nerve tissue is firing when we don't need it to, just as hay fever comes when our sneeze reflex is triggered by some antigen that's not really sneeze-worthy.
- Meme Theory—Liar because crazy. Memes are programs of cultural DNA; they replicate if and only if they force us to believe them and spread them.

In discussions like this, we are usually ready to have our beliefs challenged and to hear the experts lay down some science. However, one thing science can't do is tell us what stand to take on science as an approach to reality and the rest of our lives. Some scientists and philosophers of science will deny this and say that of course science tells us how we should approach our lives and the rest of reality. Obviously, science tells us we should do it scientifically. But when they're saying that, they're not doing science—they're doing science journalism, or maybe science advocacy.

Science doesn't tell us what we should think about science. To see how this is so, all you have to do is take any of those explanations above—Marxist, psychoanalytical, neurobiological, or meme—and apply it to itself. Thus, Marxists believe in Marxism only because it's in their class interests to do so; psychoanalysts believe in psychoanalysis only as a defense against anxiety; neuroscientists believe in neuroscience only because their brains have evolved to see causation; and meme theorists believe in memes only because the meme complex called *meme theory* has hijacked their brains and made them replicate it. These theories all explain themselves just as much as they explain Santa Claus. So it can't be the case that just because something has a supposedly scientific explanation, we should stop believing in it, or we would stop believing in scientific explanations. These theories explain themselves and they explain Santa Claus. How to go on vis-à-vis the theories or vis-à-vis Santa Claus once we realize that isn't a scientific question anymore.

You can compare the role of an intellectual theory to the role of money. A textbook in economics or finance may tell you how to go on if you want to make a lot of money, but it won't tell you how to decide how important money should be in our lives. That's a question we can debate and consider positions, all the way from making all our choices based on the financial upside to ignoring money and wandering America as hippies; or we can stake out an intermediate position. Similarly with science, we can embrace it totally, ignore it, or live our lives somewhere in between.

You might think it's obvious that if Tammi says she believes in Santa, she is crazy or lying, and back up your argument by the correct points that crazy people don't know they're crazy, and liars usually lie about whether they're lying. But there are two intertwined problems with this approach—one is ethical and the other is epistemological.

First Problem (Ethical): It is obnoxious and rude to go around accusing the parents of our kids' friends and other people of being crazy and liars. Tammi doesn't seem to be lying—she has her son's best interests at heart or at least seems to.

Second Problem (Epistemological): This is summed up in the old joke in which one Anglican priest explains the meaning of orthodoxy to another one—"My doxy is orthodoxy, your doxy is heterodoxy" ("doxy" is an old word for a prostitute).

The point of the joke is that "sane" and "truthful" need to be defined in such a way that we can tell who is crazy and lying without smuggling in our own other beliefs. Otherwise, saying

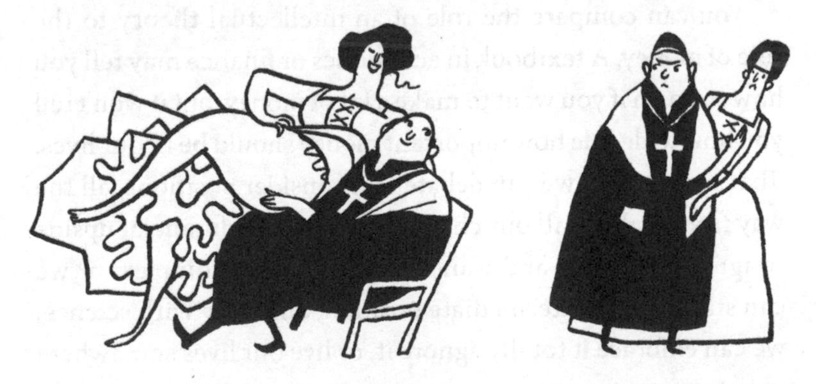

"you're crazy to believe in Santa Claus" is just like saying "Santa Claus does not exist!" in a louder, more hectoring tone of voice. It's a personal attack masquerading as a psychological explanation.

If we assume that Santa Claus doesn't exist, we might be able to argue that Tammi is crazy, but we can't use the fact that she's crazy to prove Santa doesn't exist. Maybe though there is a more direct route to proving Santa doesn't exist. If we want to know if Santa Claus exists, couldn't we just look out there and see if there is an object in the world that corresponds to my belief? But what does it mean for a belief to "correspond" to a thing? Is it a clear idea or just a fuzzy metaphor that's too murky to illuminate what exists and what doesn't? Consider the following thought experiment:

Imagine a field so big we can play the biggest game of red rover in history in it. Imagine we could open up our skull and have all the beliefs get out and stand on one side of the field, holding arms. On the other side of the field stand all the things. One by one, the beliefs call out what they are about. When the belief in Africa calls out his name, "I'm a belief in Africa!" the actual object—Africa—raises its hand, and they go off to a side field labeled TRUE BELIEFS. "Bees! I'm a belief in bees!" "Great! We are bees!" And they go off together. "I'm a belief in the planet Neptune!" "I am the planet Neptune! Let's get a drink!" And off they would go paired up. At the end of the day, a few beliefs would be left standing on their side of the field. They raise their hands: "I'm a belief in the lost continent of Atlantis!" And nothing answers on the other side. There is no lost continent of Atlantis. "I'm a belief in pixies!" No answer. There are no pixies. "I'm a belief in Santa Claus!" No answer because there is no Santa Claus. The belief in Santa Claus is wrong because there is no Santa Claus to correspond to it.

The first problem is that our beliefs don't separate themselves into little bits. How would we count beliefs? Is my belief that Africa exists a superbelief made up of beliefs in all the people, countries, and animals that I believe are in Africa? Or is it part of a larger belief that the world is divided into landmasses? Or a still larger belief that there are such things as physical objects of which Africa is an instance? All and none. My beliefs form a web or, better yet, a world. If anything corresponds to anything, it is the whole assemblage of beliefs, all linking arms, who correspond to the whole assemblage of facts, all linking arms. My mind corresponds to the world as a whole.

But there's a much more serious problem. When we imagine playing our game, we imagine that we ourselves are standing in the field somehow adjudicating the game. We are looking at the beliefs on one side of the field and the things on the other. But when we look at a thing and see it, that is just another way of saying we believe that that thing exists. There is no way to step outside ourselves and examine the world and our beliefs from the side.

Consider this classic sketch that illustrates epistemology:

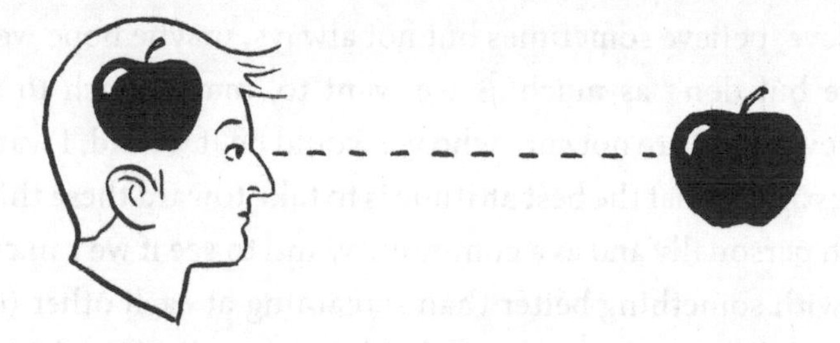

What is inside that head that looks like an apple? It's just a bunch of atoms, or if you prefer, neurons and glial cells, or if you prefer, an organ consisting of a prefrontal cortex, a cerebellum, the periaqueductal gray, the hippocampus, and so forth; but there is nothing in there that looks like an apple. And when can we stand where the picture invites us to stand—looking at the belief and the apple from the side? Never. We have to be within our beliefs, developing them.

We can't examine beliefs and apples separately and figure out if the beliefs match the apples, because the beliefs and the apples are part of a single phenomenon. They evolve together, much as flowers and bees' eyes do.

That's why I'd like to formulate the issue of Santa Claus as one of resolving an internal tension within a self, and an external tension between one self and others. It lets us get at the issue that there is something a bit funny about believing in Santa Claus without appealing to murky notions of correspondence between the content of internal beliefs and external reality.

This book is about things we are not sure we believe, half

believe, believe sometimes but not always, maybe hope we believe but don't as much as we want to, maybe wish to stop believing but are not sure who we would be if we did. I want to investigate what the best attitude is to take toward these things, both personally and as a community, and to see if we can come up with something better than screaming at each other (or at the recalcitrant parts of ourselves) "you're a liar!" and "you're crazy!" If Santa Claus is that for you, fine. If you happen not to believe in Santa Claus, maybe because some wiseass kid like my son told you that he doesn't exist, pick something you believe in but isn't universally acknowledged as real. I would suggest the point of your life and, in fact, everything. Your life will end someday, and so will everything else—given that, what is the point of doing anything at all? Some answer that it's doing what God wants, but why does His life have a point, and if it doesn't, how can his point-free life give a point to yours? Some think the point of life is to reproduce our genes, but that seems just as far-fetched. Supposing that in a distant galaxy, there was a wormhole with the feature that anything you shoot in one end would be shot out the other end duplicated a trillion trillion trillion trillion times. We wouldn't drop everything and mount an expedition to shoot a human skin cell into the wormhole, even though it would reproduce our genes better than we ever could, because—it would be pointless! So duplication of our genes is not inherently pointful, and is certainly not the point of anything else.

Some people think the point of their life comes from the fact that they freely chose it. On first blush, this idea has a certain macho charm, but if you think about it for a couple of seconds, you'll see it's problematic too. If I can give my life a point by willing it at 2:00 P.M., then I can will it to have a different point at 2:01 P.M. My life could be a succession of acts of will, but what would be the point of those acts? And what would tempt me to want one point rather than another? If I'm imposing a point on my whole life, what makes that act of imposition *mine* and not just a random event?

My point right now is not that any of these answers to the question "What's the point?" is right or wrong, but just that none of them is uncontroversial or provable. If you're like most people, I would guess you don't have a single firm answer to the question "What is the point of your life?" and that you oscillate among several, so whatever your answer is functions as your own personal Santa Claus.

PART 1

LOGIC

1

This Title Doesn't Describe What's in This Chapter

If we can't just stand outside ourselves and inspect our beliefs and the objects of the world and see if they "correspond," what should we do? Logicians argue that the minimum requirement for being a successful thinker is to avoid contradiction. That is, we shouldn't say of the same thing that it is A and not A. We shouldn't say that Mount Everest is a mountain and it is not a mountain. Why not? If we say that, we haven't successfully said anything! "Is it a mountain or not?" our listeners are entitled to ask. We have, in the literal meaning of "contradiction," spoken against ourselves. We noticed that Tammi's attitude toward Santa Claus was contradictory—she believed in him and she didn't. The path of logic views contradiction as a problem that needs solving and gives us a set of tools for thinking more clearly that can get us out of our fix.

Logic goes back at least to the sixth century B.C.E., and it seems to have sprung up simultaneously in Greece, China, and India. It makes sense that it arose in these particular civilizations and not in Egypt or Mexico because these were all places where a farmer who worshiped one god, had one set of laws, and danced a certain way would meet up and do business with a farmer who worshiped a different god, had different laws, and danced a different way. None of these civilizations at the time had a monolithic empire in charge who could just impose one city's gods, laws, and dances on everybody else. So they had to figure out a way to talk together and work together.

I'm calling it a path because logic is something that we choose to make part of our lives because we want it to do something for us. It has to be that way because logic is not the sort of thing that wants things for itself. Either logic is our tool or it is somehow part of the structure of reality, but if it is part of the structure of reality, then by all accounts it is doing just fine on it own. Logic is not irritated, sad, depressed, or disappointed in us for being illogical or irrational. It is not your mother. Logic is just hanging out, available for us if we need it. So either logic is a construct that we design for our purposes or it is an aspect of reality that we are able to tap into for our purpose, but in either case it has a purpose.

What's the purpose?

Aristotle, a philosopher whose influence on Western (and Muslim) thought was so great that for a thousand years he was simply called The Philosopher, said that the essence of logic is the law of self-contradiction. Nothing can be A and not A at the same time and in the same respect. So logic's purpose is to take us from a situation in which we are at odds with ourselves to one where we are consistent with ourselves. It is a path from a certain kind of cognitive dissociation, or disequilibrium, to one of harmony and clarity. If we state our beliefs clearly, and weed out the bad contradictory ones by means of investigation and further clarification, we will ultimately reach the goal of having no inconsistent beliefs. We will know what exists and what doesn't exist, for example. We will be self-consistent.

Here is an example of how this is supposed to work:

George is friends with Eddie, and one day Eddie shoplifts. The shopkeeper asks George if Eddie stole anything. George is torn in two directions: between loyalty to his friend on the one hand, and loyalty to what's right and wrong on the other.

Let's imagine the following dialogue between George and a logician:

Logician: You're in quite a fix, son.

George: And how, Logician! Can you help me?

Logician: Well, first let's define your problem. Do you think "People should turn in people who commit crimes" is true?

George: I sure do.

Logician: And do you think "People should protect their friends" is true?

George: Assuredly!

Logician: But since Eddie is both your friend and a person who committed a crime, you believe both "George should turn in Eddie" and "George should protect Eddie." That's a contradiction.

George: You said it! What should I do?

Logician: Define your terms and examine your premises. What is a crime?

George: Breaking the law.

Logician: Is it really true that people should always turn in people who commit crimes? What if you were in Nazi Germany and a friend of yours was breaking the law by sheltering a Jew—do you think you should turn him in?

George: No.

Logician: Then it's not true that "People should always turn in people who commit crimes." So you can protect your friend and still be a good person. Contradiction resolved!

George: Thanks, logician!

Of course this could have gone the other way. The logician could have pointed out a problem with the generalization that good people always support their friends, referencing good people who sometimes turn in their friends when the friends are bad. Or he could have applied the analytic X-ray to the concept of "support"—maybe the way good people support their friends is by turning them in when they commit crimes so they can receive help before they get in deeper trouble. What all these have in common is that they give George a path from a situation of contradiction to one of consistency.

Sometimes logic works amazingly well to resolve our problems. For example, in one of the dialogues of Plato, a sophist* is deliberately confusing a dog owner by using the following argument:

1. You have a dog.
2. Your dog has puppies.
3. If A has B, and B has C, then A has C.
4. The only animal that can have puppies is a dog.
5. You have puppies. (follows from 1, 2, 3)
6. You are a dog. (follows from 4, 5)

I should point out that the person in the dialogue is not a dog. He is a human.† But, luckily for him, the sophist's reasoning is logically fallacious. Just because you have a dog and she has puppies doesn't mean that you are a dog. Why not?

Because the word "have" is being used in two different ways: (a) "to have ownership of" and (b) "to give birth to." Once we realize that these are two different concepts that just happen to

---

* A sophist is either a philosopher who works for money, or a philosopher you don't like.

† You could have deduced that from the fact that dogs don't talk.

be expressed by the same word, we can see that this argument is unsound. The sense in which you have a dog in 1 is not the sense in which your dog has puppies in 2, and while 3 might be true of ownership in certain legal contexts, it's not true of the relationship of giving birth.

So logic saves the day.

It gets us out of some conceptual tangles.

It is awesome.

Go, logic.

Sometimes, though, logic is worse than useless. When we clean up all our statements and we know the answers to all the hard questions like "When do you betray a friend?" "When do good people commit crimes?" "What exactly does support mean?" we will still have some questions left that are impossible to resolve by means of logic. These are the so-called logical paradoxes—sentences that, logically speaking, must be both true and false. Here's one:

"This sentence is false."

If the sentence is false, then the sentence is true. If the sentence is true, then it is false. If we want to go through our beliefs and use logic to clarify them, this sentence will always be a problem. In other words, none of the techniques that helped us deal with the problem of George, Eddie, and shoplifting is available to us, because the only words in this sentence are very basic, bread-and-butter logical words.

When the Logicists tried to put math on the basis of set theory, a whole other group of paradoxes became important, the so-called set-theoretical paradoxes. Here's one:

There are a lot of different kinds of adjectives. Some of them, like "English," describe themselves. Because "English" is an English word. But others, like "long," don't describe themselves. Because "long" isn't a long word.

Let's invent an adjective—"self-describing"—to describe words that describe themselves, and another one—"non-self-describing"—to describe words that don't.

Some examples of self-describing words are "adjectival," "polysyllabic," and "consonant-containing."

Some examples of non-self-describing words are "Chinese," "incomprehensible," and "vowel-less."

Now, the paradox comes when we ask about the adjective "non-self-describing": Is it self-describing or non-self-describing? If it is self-describing, then it is non-self-describing, and if it is non-self-describing, then it is self-describing. So, as with the liar paradox, our minds kind of shuttle back and forth between two mutually opposing positions.

The solution the philosopher Bertrand Russell suggested for these paradoxes was something called the ramified theory of types. The spirit of this proposal is that Russell will give us a way of talking very carefully, so we never will get into paradoxes. Here goes:

We are not allowed to define adjectives the way we defined "self-describing." Instead, we have to be very strict about how to define an adjective, make it clear what sort of objects it applies to, and never define one in such a way that we can state a paradox. "Long," for example, is allowed to apply to objects but not to words. If you want to have an adjective that describes

words, which in turn can describe objects, it has to be a different kind of adjective. It comes from a special box of adjectives that are supposed to describe not objects but words. If you have an object and you want to make a meaningful statement about it, you are allowed to take an adjective from the box of adjectives that describes objects, and NOT an adjective from the box of adjectives that describes words.

On the other hand, if you want to make a meaningful expression about a word, you can take an adjective from the special box of adjectives that describes words, but NOT from the box of adjectives that describes objects. What if you wanted to say something about an adjective that described words? Not a problem. There's another box for that, a box of adjectives that describes adjectives that describe words. These are second-order adjectives. There are an infinite number of boxes, but Russell provides us with rules of how you can put them together. And he designs the rules so you can never formulate a paradox.

So the question "Is 'self-describing' self-describing?" is one that is impossible to ask if you follow Russell's ramified theory of types. Since there are no legitimate adjectives that can describe both words and adjectives, there are no adjectives that describe themselves. If you stick with the rules, you won't state any paradoxes. It's like an intellectual hygiene.

The logician and philosopher Alfred Tarski did something very similar with the liar paradox by defining "truth" in such a way that we would never be able to have a word "true" that could apply to a statement that had the word "true" in it.

Tarski first of all defined "true" for an object language, L, by giving rules for generating what he called T-sentences. The rules aren't important; all we need to know is that Tarski shows us how to generate the T-sentences *without* appealing to the notion of truth, and they all look like this:

"'Snow is white' is true-in-L" if and only if snow is white.

"'Grass is green' is true-in-L" if and only if grass is green.

In other words, if you want to talk about the statement "Snow is white," you can, and you can apply the predicate "True-in-L" to that statement. However, when you do so, you are not talking L anymore, you are talking the metalanguage L1. The object language is called an object language because it talks about objects. The metalanguage, L1, in contrast, talks about sentences in the object language. The metalanguage contains expressions like "True-in-L" and "False-in-L." There is also a meta-metalanguage, which talks about expressions in the metalanguage. In that language, you can say, "'"Snow is white" is true-in-L' is true-in-L1."

But in what language can you say, "This sentence is false"? No language. There's "True-in-L," "True-in-L1," "True-in-L2," and so on, but no global word "true" that could get us into trouble.

Readers whose brains are good at puzzles will have spotted the problem right away. In order for Russell and Tarski to state the problem they are addressing, and explain how they plan to address it, they have to break their own rules!

So Russell has to say something along the lines of:

"Don't ask the question 'Is "self-describing" self-describing?' That expression doesn't make any sense. Instead, use my theory of types, which makes sure that adjectives make sense only when you are very strict about what they apply to."

But the thing that Russell just said violates all his rules. He says that "Is 'self-describing' self-describing?" doesn't make any sense. But it has to make sense because he just used it! And we understood it! He didn't say:

"Don't ask the question 'Bazawagajayjay?': That expression doesn't make any sense!"

Which is uncontroversial.

He said, "Don't ask the question 'Is "self-describing" self-describing?': That question doesn't make any sense."

Which is obviously false. "Is 'self-describing' self-describing?" has to mean something or we wouldn't be able to tell it is different from "Bazawagajayjay?" and we wouldn't be able to think about how to think about it.

This is a big problem! An equally big problem is that his answer, "Don't use words in a way so that they can freely range over all of reality—be sure you circumscribe them to the type or level of reality they are allowed to describe," also breaks the rule. Because it uses the expression "words" to specify how to use them, and that expression ranges over all of reality.

Some philosophy professors I've discussed this with have tried to defend Russell and Tarski by appealing to the use/mention distinction. They argue that when Russell says, "Don't use a self-describing word," he is mentioning the word "words" but not using it. But I think this response fails, because mention is a kind of use. When we mention a word, we use it. That is, the same language-understanding part of our heads that allows us to use the word "words" in the expression "words are fun!" is used when we "mention" it in the expression "self-describing words are to be avoided." They are both uses of the word "words." In a sense, words are like kisses—if I kiss you in order to demonstrate why it's inappropriate for me to kiss you, I've still kissed you. There's not a special kind of "kiss in quotes" that we can use for making points. If you don't believe me, try it at your next sexual harassment seminar and let me know how it goes. Kisses have the power to burst out of the quotes and get real. As do words.

The claim for logic was that it would get us out of division by means of clarity, but when it comes to paradox, logic fails on its own terms. It is unable by its own terms to express either the problem or the solution clearly.

Once we notice this failure in the case of logical paradoxes, we can learn to spot it elsewhere. For example, a movement in early-twentieth-century philosophy called logical positivism tried to solve all problems, not only logical ones, by putting certain strict requirements on meaning. The positivists held that for a sentence to be meaningful, it had to be either

(a) a logical truth, such as "a = a,"

or

(b) scientifically verifiable.

They were trying to clean up thought and language by wiping off all the metaphysics and religion, because they thought metaphysics and religion were bad news. European civilization had just nearly destroyed itself in World War I, and people were understandably on the market for radical solutions. The biggest problem with logical positivism was that it was, by its own accounts, meaningless, since, clearly, the statement "The only meaningful statements are either logical truths or scientifically verifiable" is not itself either logically true or scientifically verifiable. Nobody is going to look into a test tube full of sodium or up a beaver's anus and report, "I just discovered that all meaningful statements are either logical truths or scientifically verifiable."

So logical positivism is either false or meaningless. Probably false. Sometimes, as a work-around, the logical positivists compared their own writing to art. They said that while meaningful language uses arguments to convince people, art just has an emotional effect on them. So the statement "All meaningful statements are either logical truths or scientifically verifiable" was not meaningful, but it could have an effect, like the lyrics to "I Am the Walrus." The obvious problem with the statement "Logical positivism is art" is that it fails as art.

A sophisticated version of the logical positivist program is laid out by Ludwig Wittgenstein in his *Tractatus Logico-Philosophicus.* Wittgenstein puts forward a theory of meaning similar to that of the positivists, namely that a sentence is meaningful if it directs our attention to a bunch of different possible states of affairs, and picks out one of them as being the way things are. Then Wittgenstein goes on to address the problem the logical positivists foundered on: What is the status of his own sentences? If the only sentences that are true are ones that pick out a particular state of affairs as being true, what about the sentence "The only sentences that are true are those that pick out a particular state of affairs as being true"? Wittgenstein has the courage of his convictions and says that this sentence is nonsense.

"My propositions serve as elucidations in the following way: anyone who understands me eventually recognizes them as nonsensical, when he has used them—as steps—to climb beyond them. (He must, so to speak, throw away the ladder after he has climbed up it.)" (Wittgenstein 89)*

---

* Page or (in this case) section numbers will refer to the edition listed in "Suggestions for Further Reading" at the back of this book.

Wittgenstein was very, very smart, but this statement is very, very stupid, as anybody would know who has ever used a ladder to climb up somewhere. If his propositions are a ladder for helping us climb somewhere and we throw it away, we will be stuck! Why would we throw them away? What if we decide where we go with them isn't so great? Or we remember when we're up there that we forgot something? Or we like it up there sometimes but not all the time? We could be up there and it could get very rainy, for example, and we might want to come down until it blows over. Why would Wittgenstein advise us to climb up a ladder and then throw the ladder away? It's terrible advice!

We started our discussion of the ontological status of Santa Claus by noticing that the mind seems to oscillate between two points of view—that Santa is real and that he isn't. Does the path of logic help? Maybe a way to resolve contradictions is the method that takes as its watchword "The Principle of Non-Self-Contradiction." But—too bad!—in the purest recesses of logic, when logicians are talking about logic itself, they say deeply illogical things. They say that their own opinions about the relationship of logic and life are both true and nonsensical. The mind oscillates just as much in the lofty precincts of logic as it does down here on Earth! Logicians are split just like regular people! One part of the logical positivist says that his statements are important, and one part says that they are meaningless. One part of Wittgenstein says that the *Tractatus* is important and clear and one part that it is nonsense. He is as confused as Tammi!

Are these logicians crazy? Are they lying to us, saying something that doesn't make sense in order to get cushy jobs teaching logic and invitations to fancy logic parties or perhaps jobs putting on glitzy Stage Logic shows in Las Vegas?

Or, maybe, logic is just poorly equipped to deal with logical paradox and with the paradoxical situation of reflecting on its own activity. Isn't that okay?

Well, no, it's not okay, for two reasons. One is that the explicit goal of the logician is better living through consistency, and if logic is itself an inconsistent practice, then it doesn't justify itself. If I'm a rock star, it's fine that I'm inconsistent, because my claim to fame is that I make awesome music, and teenage girls want to have sex with me. But if I'm a logician, my claim to the public's attention, toleration, and financial support is that I'm logical.

The second issue is that we have made precisely zero progress in our task of resolving what our correct attitude toward Santa Claus should be.

Put it this way:

We started by noticing that we were of two minds about Santa Claus: We were tempted to believe he was real but also felt he wasn't. When we pursued a rigorously anti-Clausian position, we found ourselves in a situation where we had to embrace a theory of meaning that we also felt was meaningless. Logic has the same status as Santa Claus—we think it's real with one part of our mind and think it isn't real with another.

2

The Science of Not Being a Chump

There's a saying in jurisprudence that hard cases make bad law. Maybe "This sentence is a lie" is just one of those weird things you can't think about too much, because it makes your head hurt. We're not like robots in *Star Trek* whose heads literally explode, but it's still uncomfortable. Admittedly, this is an odd complaint coming from philosophers—isn't living with thoughts that make other people's heads hurt their patented brand of machismo?—but let's take a beat to see if we can make a case for rationality and logic that avoids the pitfalls we ran into in the last chapter.

Maybe logic and rationality are not primarily an issue of getting our beliefs in order. Maybe they are a question of getting *our lives* in order. Maybe rationality is not a standard to apply first and foremost to our thoughts; it's a standard we apply primarily to our actions, and only derivatively to our thoughts.

Maybe what makes bad beliefs bad is that, ultimately, they will lead to foolish or self-defeating actions. So the problem with believing in Santa Claus is not just that it's a bad thing to think—although it is—the real problem with believing in Santa Claus is that it is a bad thing *to do*. Don't believe in Santa Claus, in other words, because it's going to get you into trouble. Not just conceptual trouble or "philosophers will think you're dumb" trouble either. Real trouble. You'll lose money.

The best candidate for cashing this idea out is the notion of practical rationality. The crude version of this idea is that, if you want something, you should figure out a way to get it. And if you want a whole lot of different things, you should make sure that your beliefs are correct, so you can get those things, and also make sure that your approach to life isn't so incoherent that you can never be satisfied. So, for example, if you prefer a piece of pizza to a can of soda, you should not also prefer a can of soda to a piece of pizza. Otherwise, you will find yourself the unwitting actor in the following sad little play.

INCONSISTENT PREFERENCES:
A TRAGEDY IN ONE ACT

Characters: You, Me.

OPEN ON YOU, with a can of soda and a hundred dollars. Enter ME.

Me: You prefer a slice of pizza to a can of soda, right?

You: Yup!

Me: Great. Pay me a dollar and I'll trade you your can of soda for this slice of pizza.

You: Okay!

You trade the soda for the pizza and now have one slice of pizza and ninety-nine dollars.

You: Thanks!

Me: Any time! But—you also prefer a can of soda to a slice of pizza!

You: I sure do! That stuff's great!

Me: Great! I happen to have a can of soda!

You: The one I just gave you!

Me: The same! How'd you like to trade it for a slice of pizza and one dollar?

You: I would! And in fact I have one—

Me: The one I just gave you!

You: Yes!

You trade the pizza for the soda and now have one can of soda and ninety-eight dollars.

CUT TO

Ninety-eight trades later. You sit there with a can of soda and no dollars, crying.

You: Boo-hoo! Boo-hoo!

Me: What's wrong?

You: I really want that slice of pizza, but I don't have any money.

Me: I can solve your problem.

You: Great! How?

Me: I'm going to let you apply for a credit card.

CURTAIN

It's a sad play because your preferences are incoherent, and that allowed a tricky person (me) to get a naive person (you) to bleed yourself dry. So it seems intuitively correct that there's something wrong with having preferences like that. It's irrational. Whatever the preferences are, it's bad to have them be nontransitive, in much the same way that whatever your beliefs are, it's bad for them to be contradictory. You can lose money, you can lose opportunities, you can waste your fertile eggs. Anything that is of value, or anything that you prefer over anything else, can be wasted if you are irrational, and maximized if you are rational.

It may seem like this is cute, interesting, smart, and a complete waste of time, because nobody could actually prefer Coke to pizza and also pizza to Coke. However, the theorist of practical rationality will point out that there are more sophisticated ways to be irrational. For instance, there are more complicated ways to have inconsistent preferences: You could prefer a can of soda to a tuna fish sandwich, a tuna fish sandwich to a slice of pizza, and a slice of pizza to a can of soda with results similar to those discussed above. And if your probability judgments are inconsistent in a certain way, a "cunning bettor" can make a "Dutch book" against you: a series of gambles that you will always lose. If you are playing the game of life in such a way that you will always lose no matter what happens, then you are irrational.

In the life cycle of a theory, it starts off simple and then gets fancier and fancier, as brainy thinkers mount objections and the theory's proponents develop a more subtle, complex, and well-defended theory to stave them off. Then it dies. Actually, before it dies, it lives in a special preserve for theories too complicated to survive in the wild, called a university. The theory of practical rationality has passed through the same life cycle. It now exists in a sophisticated version known as the theory of subjective expected utility. This states that we should have a degree to which we like every possible state of the world, and a belief about the likelihood that anything we do will bring about that state of the world, and if we are rational, we should act so as to make the product of these two numbers as big as we can. This is called maximizing our subjective expected utility.

How does that work? Well, if there is something we think is the worst thing possible, and there's a course of action that costs us nothing to avert it, we should take that course of action. If you don't want the world to be destroyed and you can get the world to not be destroyed by moving your little finger, and it doesn't cost you anything to move your little finger, move away!

If we like getting oranges and there are two buttons, one of which gives us an orange and the other one gives us two oranges, and it costs us the same to hit the first as the second button, all things being equal, we should hit the second button. Unless, of course, there's something we could be doing other than pressing buttons that is likely to bring about what it is we want. Looping back around to Santa Claus, you might say that, as a grown-up, you should not believe Santa Claus exists because the actions that go with that belief—leaving out stockings, for example—are less likely to get you what you want than other actions. In other words, they are inefficient. The idea that belief in Santa Claus is a deficiency in practical reason meshes nicely with our sense that those who say they believe in Santa Claus are either crazy or lying, because those both seem to be deficiencies, not in cognition *simpliciter,* but in how we conduct our lives.

But is practical rationality the best way to conduct our lives? The most beloved Christmas stories suggest the answer is no.

Take O. Henry's story "The Gift of the Magi." The story tells us about two newlyweds, a young woman whose most prized possession is her long hair, and a young man whose favorite item is his beautiful pocket watch. They're poor, but each wants to give the other a Christmas gift. On Christmas morning, the husband presents his wife with a gift—an expensive and beautiful comb for her hair. She asks him how he could afford it. He says he sold his watch. She reveals that her gift for him is a fob for his watch. He asks her how she could have afforded *that*. And she says she got it by selling her hair.

Do we feel good about the prospects of the relationship? I think we do—just judging by my own emotional reaction, I feel I would be happy to be one of the lovers in O. Henry's story. Even though neither got what they wanted—in fact, what they got was precisely useless, namely a comb for a bald head and a fob for a nonexistent watch. Would they have done better if they had been more rational? Well, if they wanted to be perfectly rational, they could have just asked each other what they wanted and then, using their joint resources, bought it together. Or better yet, they could have given each other money. Or best of all, since money is fungible, each person in the relationship could have kept his or her own money and bought him- or herself exactly what he or she wanted! So the most rational gift for me to give my wife would be for me to keep my money and buy something for myself and to let her keep her money and buy something for herself. But that would not be a gift anymore! It would, in fact, be what we call a purchase. In fact, if you look at what makes it great to give and receive a gift—that it's surprising, that we didn't ask for it, that it affirms a relationship that is bigger than our private, selfish desires—none of it is captured by our theory of practical rationality.

If we look at Christmas carefully, it almost seems as if the whole point of the holiday is to warn us away from practical rationality. First of all, it's about Jesus Christ, who was, the story goes,* a gift from God to humanity, and was therefore both a departure from practical rationality on the part of God and, given that he sacrificed his life, also a departure from practical rationality on the part of Jesus. (I'm not a Christian, so forgive me if I express this ham-fistedly.) Eighteen centuries and change later, in "A Christmas Carol," Ebenezer Scrooge starts in a place where he believes giving money to the poor is wrong because it will simply encourage them to live when the rational thing for them to do is to die. But in the course of the story, he learns from three ghosts that a life lived according to practical rationality is actually the life of a miserable shade, sucked dry of substance, chained to a cash box. Jesus teaches us that the way to gain ourselves is to lose ourselves, Scrooge learns that the best way to live a happy life is not to figure out the best way to live a happy life, the couple in "The Gift of the Magi" learn that the best gift is what neither of them wants. None of them is living a life of practical rationality.

In our discussion of theoretical logic, I argued that our

---

* In the Bible.

minds are fundamentally at odds with themselves about certain things we care about, and that logic doesn't solve the problem. It just formalizes it. When we intellectualize a problem, we lose a lot of potentially helpful information coming in from our nonintellectual faculties: our body, our imagination, and our emotions.

These are just stories, though. They appeal to our imagination and our emotions, but maybe our job as hardheaded rationalists is to ignore them, just as we ignore the appeals made by ads. Ads tell us attractive stories about people who get the right deodorant and are mobbed by gorgeous women, but smart people who have been educated by *Mad* magazine know to resist them. And that would be a good idea if practical rationality were a reliable path we could plunk our feet on. However, just as theoretical reason has its paradoxes—trapdoors that drop you into a whirlpool of confusion—so too does practical reason.

One such paradox is Newcomb's, popularized by my fellow Brooklynite Robert Nozick in the 1960s. Here's how it goes down. In Newcomb's paradox, a wealthy, eccentric psychiatrist—let's call him Dr. Money—invites you to participate in a potentially very lucrative psychological experiment at his psychology lab/château. On Day One, Dr. Money subjects you to very accurate psychological tests: questionnaires, brain scans, and the like. On Day Two, he leads you to the door of a room that contains two identical, sealed boxes. Before you enter the room, you must announce a choice: Either you will open one box or you will open both. Dr. Money determined the contents of the set of boxes on the evening of Day One. Nothing you say will cause what is already in the boxes to be changed. The boxes both contain money. You get to keep whatever is in the box or boxes you open.

It seems on the face of it that your course of action is simple—choose to open both boxes, and collect the cash prize in each box. But there is a wrinkle that makes this setup a paradox.

The tests that Dr. Money performed on you yesterday, on Day One, were tests about whether people are the sort of people who open one box or two on Day Two, and they are *one hundred percent accurate*. Based on what he learned about you, Money did the following: If he judged you to be the kind of person who will open both boxes, he placed a crisp ten-dollar bill in each box. If he judged you to be the kind of person who will choose to open only one box at random, he placed one TRILLION dollars hard money cash in one box and absolutely nothing in the other one.

Now what do you do? Deals for lottery tickets have a value based on the size of the payoff and the likelihood of getting it. So a one-in-two chance of winning ten dollars is worth what? Five dollars. A one-in-ten chance of winning a hundred-dollar lottery is worth? Right, ten dollars. And a one-in-two chance of winning a trillion dollars is worth precisely five hundred billion dollars—the cost of beer consumption for the entire country of Ireland for 250 years! That's way more than the assured gain of twenty bucks you get if you open both boxes. The money is already there—it was placed there yesterday and nothing you do today can change what happened. The following lines of reasoning both seem to be valid:

(a) Declare your choice to open only one box. If you declare that choice, it is a fact that you are the kind of person who would open only one box. You know that Dr. Money had an accurate idea of what kind of person you are, therefore he knew what kind of person you were yesterday, and therefore by choosing to open one box, you have the deal worth five hundred billion dollars.

(b) Declare your choice to open two boxes, because whatever is in the two boxes is already there and nothing you do now can change the past. Since it's already there, you should take the course of action that is better, and two boxes always have more money than just one box.

This paradox seems to result from two equally valid ways of looking at your own choice—as free, which points toward choice b, and as pre-known by the psychiatrist, which points to a.

The mind seems to bounce back and forth between the two ways of looking at choice.

When we were looking at logic, we ran into the problem of a paradox where the mind shuttled back and forth between two incompatible ways of looking at a situation. Newcomb's paradox is putting us in the teeth of two incompatible ways of looking at a decision.

We can generate other paradoxes along the same lines:

If you know you always lose half the weight that you want to, how much weight should you want to lose, assuming you want to lose ten pounds? In this paradox, the mind shuttles between the following choices:

(a) Twenty pounds. You know you always succeed in getting halfway to your goal, so you should want something twice what you want.

(b) Ten pounds—because that's what you want to lose!

A related issue is that of self-defeating and self-fulfilling prophecies. People pay a lot of money to motivational coaches to tell them, "You can do it!" Why? Because it's true that if you think you can do it, you'll have a better chance at doing it. Let's say you are trying to save up money so you can afford to spend a weekend with a motivational speaker. Which is correct?:

(a) You should believe you can save the money, because that way you'll succeed in saving the money, and you'll be able to go to the motivational speaker.

(b) You should believe you can't save the money, because if you believe you can save the money, you won't go to the motivational speaker, because you'll think you don't need to.

or:

(c) You should believe you can save the money, because if it works, you'll have discovered that you don't need to go to a motivational speaker at all—you can motivate yourself.

It's hard to say! Does that mean practical rationality is a bad idea? Well, maybe, although these are unusual circumstances. It might be that on the whole, we can stick to rational choice as a way of living our lives and just avoid the tricky issues.

Well, okay. So what are some of the tricky issues we want to avoid?

One is gifts. We noticed when we were talking about "The Gift of the Magi" that a gift is not simply the act of transferring a certain economic asset from one party to another—that's just a payment. That's tricky because: What are we supposed to do with that thought? If we plan on giving a gift that is not wanted, then that's just a bad gift. It's the thought that counts, but if we go shopping and think to ourselves, "It doesn't matter what I get as long as I have the right thought," we, *by that very fact*, no longer have the right thought! We can be like the redeemed Scrooge, the couple in "The Gift of the Magi," or Jesus Christ only if we give up trying to be. When we try to think rationally about gift giving, we get into another kind of oscillating, bouncing mind—we think we shouldn't think about it, and we shouldn't think *that* either.

It's so annoying that I might want you to forgive me for annoying you! And speaking of that, forgiveness is also a tricky thing to think about using practical rationality. If I break your toe and ask for forgiveness and you give it, it's a beautiful thing. If I think to myself, "I'm going to break your toe because all I have to do is ask you for forgiveness afterward and I'll get it," and then I proceed to do so, the whole thing goes off the rails—you are going to feel manipulated and not want to forgive. But what if I know you're a very forgiving person, and forgiveness is always an option? I have to somehow force myself to forget that fact so I don't take advantage of you! Otherwise, if I get pushed into a tight corner from which the only egress is toe breaking, I'm always going to choose to break yours rather than your unforgiving brother's, and that's not fair at all!

If I think too much about the potential for forgiveness, I lose the beautiful part of being one of a group of people who are able to forgive one another. If I try to maximize my utility by thinking who will forgive me and who won't, I actually hurt myself.

This paradox comes up in reconciliation between enemies. It would be great for me to put down my gun, whether it be a

literal one or a gun of propaganda directed against my own people. If I respond genuinely to a gesture of goodwill on the part of my adversary, then I might disarm myself. But if I think your gesture of goodwill is calculated to get me to put down my gun so you can proceed to shoot me, then I won't put it down, even though both of us long for peace.

So gift giving is hard to understand. Forgiveness is hard to understand. So is a certain kind of openness to life. I work in Hollywood, and the following scenario comes up with some frequency: An open, vulnerable, sincere creative person who likes singing or joking or telling stories decides to make a living at it, and moves out to L.A., whereupon he or she is lied to and ripped off by the next fifty people he or she meets. At this point, the person becomes guarded, and it's very hard to be creative yourself and to forge partnerships with other people if you're guarded, because what makes you an artist in the first place is the willingness to be open and vulnerable. That's why people should work with their friends from high school or with me, because I'm nice. But failing that, you are in a bind where you want to be naive in order to be unguarded and creative but can't get there by deciding to be that way, because that's not naive at all.

Another tricky, hard-to-think-about-without-oscillating thing is emotion. There are a lot of reasons why I might want to feel and express genuine emotion. It helps me have relationships with other people. It feels great. It's good for my health. But I can't get to genuine emotion by calculating it, because, by definition, genuine passion is uncalculated and unfeigned. If I read in a health magazine that it's good to have three spontaneous belly laughs a day, and sit down and try to induce them, it won't work, because I'm being too serious. We can't get ourselves to laugh by planning to. A life of pure calculation will be a dead life without the pleasure of laughter or, for that matter, the relief of tears.

An interesting theory holds that one reason anger evolved is that it lets us get out of the following dilemma of rational choice. Suppose I am being bullied by someone stronger than me. For no reason at all, let's call him Jimmy Toscano. If we're both rational and we both know it, Jimmy Toscano has no reason to fear me. Jimmy Toscano knows that I know that he's stronger than me, and therefore if it comes to a fight, I will likely get beaten, so, being rational, I will not fight. It's bad news for me if a bully like Jimmy Toscano knows that I will never put up a fight. I will never get to eat my lunch. However, if I'm angry enough, I will stop thinking and will just attack him, consequences be damned. So if he knows I'm capable of anger, he will have to control his bullying more than he would if he were dealing with a hyperrational little nerd. However, for this to work, this kind of anger has to be real. If Jimmy Toscano knows that I know that it's a good idea to be angry because it will make him take me more seriously (perhaps because he saw me reading an article on the evolutionary biology of anger [and had someone explain it to him]), then he will take me and my book on anger and cram us both in a garbage can. I will remain an easy-to-pick-on nerd, but one who has a theory of anger. I need to really get angry, and not get angry because I think it's a good idea.

Another idea closely related to passion is commitment, because passion takes us outside of ourselves and causes us to form committed relationships—with another person or with a group. It also feels amazing to be swept along by an emotional tornado. But I can't get to passion and commitment by deciding that I want some and putting them on my plate, like bacon and eggs at the all-you-can-eat buffet at the Hilton, because the recipient of my commitment will be rightfully worried that what I bestow for a reason I can un-bestow when the reason changes. Reasons and thoughts are a lot easier to change than commitments. Even if commitment is something I can think myself into, it still shouldn't be something I do think myself into, or it's not a commitment—it's a fake. Which is the worst! It's like writing an insincere love letter.

This, by the by, hooks back to our discussion of openness to life and creativity. Creativity is a matter of a passionate commitment to myself, or maybe to my own ideas or the flow of experience that passes through me. As long as we stick to copying something that has worked before, we are not being creative: We are plagiarizing our former selves. Therefore, if we calculate that something will be a good creative move, that our creative production could use a little more of this or a little less of that, we are, by definition, no longer being creative. I don't want to just be creative in my work as a writer, I also want to be fresh, creative, and spontaneous in my life. But if I think to myself, "Let's be spontaneous right now," then I am, by definition, no longer being spontaneous.

Another idea that makes the mind bounce is faith. Let's say I would like to believe in God (or the future just society [or the ultimate goodness of my fellow man]) because I'm tired of being depressed by a meaningless universe (or injustice [or how crappy people are]). I even believe that if I have faith, I will move mountains and do wonders—because I won't be constantly doubting my actions. If I believe that the reason I believe in God is that I would like to, and not that He exists, then I don't believe in Him. I have to believe that the reason I believe in Him is that He exists.

In our lives, we sometimes grow by adding a little bit more to what we had before. We were good enough at woodworking to make a shelf, and then we got good enough to make a chair. Sometimes, though, we grow by making a radical shift—we were good at woodworking and then we realized that all our skill at woodworking was a hollow triumph, and we ran away to Ethiopia to serve the poor.

We had our life figured out when we were kids and then bam! puberty came along and totally upended what was important to us. We had our lives figured out as a couple and bam! we had a kid and everything went kablooey. As a culture, we were just fine doing Newtonian physics and then boom! Einstein, in Thomas Kuhn's contribution to the T-shirt, "subverted

the dominant paradigm." I was trapped in my little apartment arranging my dolls, and then you came along and set me free.

Therapy, education, and paradigm shifts don't promise just to give us what we want, they promise to change what we want and reorient our priorities. We may plan for them, but what we plan for is not what we get. We perform a bait-and-switch on ourselves, promising ourselves gradual change but giving ourselves a revolution. It's only postrevolution that the new person we are can look back and make sense of it.

Eroticism is another nice thing. Imagine that two rational people, call them Richie Rosenstock and Anna Maria Leslie Kolokova, are coworkers. He is a temp installing a new e-mail system, and she is assistant to the vice president of HR. Richie and Anna Maria Leslie are attracted to each other, and they'd like something to happen between them. Imagine that they brush against each other in the copy room, and this is potentially the big moment for them to stop being coworkers and start being something more. But imagine further that each of them thinks, "It would be great for this to be the moment when this turns erotic. This brushing of our bodies against each other in the copy room is exactly the sort of thing that can turn coworkers into lovers." It's not going to work for Richie and Anna Maria Leslie because that's a very un-erotic thought! And that's not just a happenstance to how we're made or a peculiarity of our culture. Part of what makes something erotic is that it is opening myself to the unknown. The moment I think brushing against my coworker in the copy room is exactly the sort of thing that would get this relationship erotic—brush! brush!—it stops being flirtatious and becomes gross and methodical! Potential lovers would like the ambiguous nature of their relationship to flower naturally into something erotic. The more they ask each other "What exactly is happening here? Is it sexy yet? How about now? How about now?" the less chance there is that anything will happen.

Childish innocence is great—it's sweet and refreshing and puts a smile on your face. But an adult pretending to be a child because he'd like to be innocent is macabre and creepy and makes you want to throw up.

We know that the most blissful moments in our lives are those when we are swept up so deeply by an experience that it would never occur to us to check our e-mail. If we are fully in the moment, we wholeheartedly care about one thing and anxiety drops away. Wholeheartedness is great, but obviously we cannot be wholehearted if we are thinking both "I like this baseball game" and "It's great that I'm being wholehearted." Then we have at least two thoughts.

So if you don't want to be a chump, you can follow a life of practical rationality. You won't have creativity, or eroticism, or faith, or passion, or be able to give or receive forgiveness, or be committed to anything larger than yourself, or be spontaneous or wholehearted, or ever feel fully alive. But at least you won't be a chump.

3

Is the Smartest Thing to Make Yourself Stupid?

Obviously, there's a problem here. Just as logic seemed to turn around and, Ouroboros-like,* bite its own tail and disappear, so rational choice, the theory that life is about making careful and prudent bets, proves itself to be a bad pony to bet on. The theory of subjective expected utility fails as both description and norm: It's neither an accurate description of how successful people live their lives nor good advice for us to follow if we want to be successful.

The paradoxes of theoretical reason resemble those of practical reason. If someone tells you, "This sentence is a lie," your mind will shuttle back and forth between thinking that it's true and thinking that it's false. If somebody tells you, "The only way

* Have you read *The Worm Ouroboros* by E. R. Eddison? If you like high fantasy or worms ouroboros, you should.

to be happy is not to try," your mind will shuttle back and forth between not-trying and trying, at least insofar as you need to evaluate your not-trying to see if it's working or not. Tarski and Russell tried to neutralize the paradoxes of the intellect by advising us to distinguish thought from thought about thought. In a similar way, theorists of rational choice have tried to deal with these paradoxes by bringing in concepts that would explain how, sometimes, it is rational to not be so rational. For example, Jon Elster appeals to the concept of the "intrinsic side benefit." If you're an insomniac who wants sleep, you can't get to sleep by trying to get to sleep. That will just keep you awake. Instead, you should recognize that sleep is an intrinsic side benefit, and try to do something else, say count sheep. You don't really want to know how many sheep there are, but by trying to count them, you are doing something that you know puts you in the path of sleep and thereby allows sleep to come.

The psychology of the rational person trying to be irrational is like the psychology of the person trying to avoid a paradox who uses the logical systems of Russell or Tarski. Just as, on some level, Tarski has to understand the statement "This sentence is false" in order to formulate a philosophy that avoids it, so on some level, the rational choice theorist trying to become spontaneous has to understand that spontaneity is an intrinsic side benefit and that he is trying not to try to get it. To try, I have to not try. But if my trying by means of not-trying is to succeed, I need to evaluate my success periodically. I need to see how close I've come to spontaneity, and if I'm not making any progress, make mid-course corrections. I have to split into two parts: the part that knows the rational thing to do is not to try to be spontaneous, and the part that has to try to forget that and just be spontaneous!

The rational theory of irrationality splits us on the social level as well.

There are many good things that have the following structure: As a group, we all want the good thing, but as individuals, if each of us thinks the good thing is something he or she is trying to get, we all lose it. For example, sharing. The tragedy of the commons was an event in the English social history of the 1700s that followed a paradoxical logic. English farmers prospered from the use of "the commons"—a big meadow so called because nobody owned it; everybody could use it to graze his animals in common. However, each farmer could reason that if he enclosed the little bit of the commons that was adjacent to his farm, he would get the benefit of this new private lot and also the benefit of the commons. The problem was that everybody reasoned this way. Soon enough, there was no commons left, and everybody was worse off. Just like the person aiming for spontaneity who spoils his spontaneity by trying to achieve it, the farmers are rational, with the result that they lost everything.

How could you solve the problem? One option is benevolent conspiracy. You and your friends could form a small cabal of

wise philosophers* who lie to the people, telling them that the commons are ruled by an invisible fairy named Commonus. Commonus flies around on a magic dormouse, wears a crown of hawthorn berries, and does one other thing—oh, yes, if any farmer encroaches on the commons, Commonus gives him fatal eyeball cancer.

The farmer who is forced by the logic of his own rationality to fence in the commons abstains from doing so because he fears Commonus. Everybody wins at the cost of believing something you and your friends made up.

---

* Can we have a clubhouse? Yes, you can have a clubhouse. Can we have a special secret hat that looks like a regular hat but has a weird secret flap that folds out and says BENEVOLENT PHILOSOPHER CABAL in code in gold thread? Yes. Yes you can.

We're not exactly faced with the tragedy of the commons, but in the nineteenth century, we were faced with a different sort of environmental threat. The practices of un-self-conscious and spontaneous human relations that knit individuals into a web of reciprocal generosity were under threat from a new order of life that was rationalized and explicit. This wasn't just a fear of conservatives. Karl Marx, no conservative, pointed out in *The Communist Manifesto* that everything solid was in danger of melting into air as capitalism took human relationships based on feeling and reciprocity and replaced them with relationships based on money. The wise heads in nineteenth-century New York wanted to develop a practice that would stand up against the cultural pressure to view all of life as explicit and transactional, something like the myth of the commons-defending Commonus. As the historian Stephen Nissenbaum argues in *The Battle for Christmas,* a conspiracy of rich Dutch New Yorkers, including Washington Irving and Clement Clarke Moore, the author of "The Night before Christmas," invented just such a practice. In the early 1800s, they worried specifically that the traditional folksy city of New York was becoming anonymous and commercial. To keep this from happening, they invented the belief in Santa Claus. They didn't invent him from whole cloth, like our hypothetical Commonus. Instead, they took a collection of Northern European legends, including Kris Kringle and Knecht Ruprecht, mixed in the carnivalesque practice of poor apprentices invading rich people's houses at Christmastime and singing until they got food and beer as a bribe to go away, refocused it on the home and the family, and we got the modern practice of Santa.

Now, whether or not Nissenbaum's right, we can ask ourselves if we would want to embrace such a cultural practice.

Let's look at some of the features of the Santa myth and how they could figure in a conspiracy aimed at helping us overcome the bad consequences of rationality.

Santa is old. The fear of the debility of old age is an anxiety at the heart of family life, which depends on an intergenerational contract: Children trust their parents to provide safety, and parents trust their children to take care of them when they get old. When we hear about families that mistreat their elderly, we are horrified. You can view intergenerational families as something like the commons: We take care of our children when they are young and weak, and they take care of us when we are old and infirm. Santa Claus is an old man, but he is very strong—miraculously so. Santa Claus embodies trust in a long life well led and soothes our fear of old age and death, which in turn makes family life possible. For our farmer ancestors, the winter solstice was a matter of life and death—as the shortest day of the year, how much food they had stored by about then let them know if they would live or starve. We're not them, but seasonal affective disorder is real, and Santa comes on the literal darkest day, bringing hope.

He also brings gifts, and as we saw above, one of the trickiest issues for the theory of subjective expected utility to deal with was gift giving. If I give myself a gift, it's not a gift. If you tell me exactly what you want as a gift, and I give it to you, it's not a gift.

If I give you a gift explicitly in return for something you did for me, it's not a gift. If I give a gift to you to cement our relationship, it's not a gift—it's a bribe. If we believe Santa gives the gifts, we don't have to worry that gift giving will collapse as a practice and we'll end up living in a world where we just tell each other what we want and order it off the Internet. Belief in Santa cloaks the practice of gift giving in mystery, so it doesn't dissolve under the scrutinizing glare of practical rationality.

If Santa Claus is a practice designed to keep a benevolent conspiracy alive, we would want to include safeguards that prevent people from thinking about it too hard. Just as counting sheep brings sleep if you don't think about how you're doing it to bring sleep, so belief in Santa Claus can give you a culture of spontaneous generosity and cozy domestic harmony, just so long as you don't think about it. The clearly irrational aspects of Santa Claus—his flying reindeer, his ability simultaneously to visit everyone's house on Christmas Eve despite the obvious challenges to our ordinary understanding of time and space, those weird elves—all serve as a warning sign, telling the rational mind to keep out. Obviously, we can't say rationally, "Don't think about Santa Claus," because that would tip off the mind that something fishy was afoot. Instead, we signify it metaphorically by the fact that Santa comes into the house at night when we are sleeping. When our rational minds are at rest, and our calculating prefrontal cortices are down for the count, Santa enters the house, the heart of our domestic meaning, and he enters through the chimney, the central spine of the house's warmth. We give him the best we have to offer our children—milk and cookies—and he gives us gifts. And, I'm given to understand, plums.

If we assume that what I have said about the limitations of rational choice are correct, and that the belief in Santa Claus would, in fact, be a good way to train ourselves into honoring and preserving all the parts of life we value that are not about rational choice, which of the following would best describe the situation?:

(a) Santa Claus doesn't exist, but we should believe that he does.

(b) Santa Claus exists.

It's a tough question, and one that seems to take us out of the simplistic approach of the LIAR and CRAZY models of Santa belief.

Imagine you had a mind-control helmet. You have the ability to make anybody have whatever belief you want. It's sort of a remote-control brain surgery device—it works by releasing a cloud of nano-drones that go into the brain, find the belief, and snip! snip! snip! replace it with a better belief. Suppose I have convinced you that the theory of subjective expected utility leads to all sorts of bad results—loss of innocence, loss of faith, clumsy flirtation, lack of cooperation, the despoliation of the environment, etc. etc. etc. Suppose, further, you buy the idea that belief in Santa Claus lets us avoid these negative effects. So you use the mind-control helmet on everybody

on Earth and convince them all that Santa Claus exists. Now December 24 rolls around. Everybody is eagerly awaiting Santa Claus, except you. What do you do?

Do you use the helmet on yourself? If you don't, you will feel a bit alienated come Christmas morn. On the other hand, it seems both icky, and a form of intellectual suicide, to operate on your own brain. So let's say you don't. You are the only one on Earth who doesn't believe in Santa Claus. Imagine, now, a friend comes to visit you on Christmas Eve.

Friend: Hey, you. Why so glum?

You: No reason.

Friend: You excited about Santa?

You: (*lying*) Sure.

Friend: No, you're not.

You: Sure I am! Fa la la! Jolly Saint Nick! Deck the halls!

Friend: Look, I know you don't believe in Santa. You believe those ancient, discredited anti-Santa-ist thinkers who think that we actually buy and wrap the presents ourselves.

You: Me? Nah.

Friend: Sure you do. You think the whole idea that Santa's sack emits a mind-control ray that makes us think we bought and wrapped the presents ourselves is far-fetched. I know because I hacked your e-mails. In fact, I know that you think the reason everybody else believes in Santa Claus is that you brainwashed us with a helmet.

You: Well . . . okay. Look, I do think that because I did! Don't tell anybody—I don't want to rock their beliefs. I did it for a good reason! Don't be angry at me!

Friend: Not a bit of it! Because I have a different explanation.

You: What do you mean a different explanation? I built the helmet! It's under my bed! I can show it to you!

Friend: I know all that, but I still believe in Santa Claus.

You look at him with amazement. The Christmas ornament you have been fiddling with drops from your hand and shatters on the floor.

You: How can you?

Friend: Everybody on Earth has a disease called overrationalism. It probably developed sometime during the Industrial Revolution, as life became more complicated and we all were required to think too much to deal with life. One of the symptoms of this disease is an inability to access spiritual realities, one of which is Santa Claus. Your helmet was the cure.

You: (*getting annoyed*) No. That's not what happened. I built it. I know what it was. It was something to brainwash people into believing in Santa Claus. You guys are all brainwashed. I brainwashed you.

Friend: I can see how it looks that way to you. But, in fact, you are like a blind man who invented a cure for blindness. Since you are blind, you have no way to understand the marvelous, positive effects of your own invention. But if you simply put it on your head and turn it on, you will immediately see what I mean! In fact, let me help you.

You: No! No! No!

You fall to the floor wrestling, bump into the Christmas tree, and knock it over. The kids, hearing the commotion, come running in.

Kids: It's Santa! It's Santa!

CURTAIN

How could you decide who is correct? Can brain science tell you? Not really! Suppose that what your helmet has done is allow a bunch of neurons in the imagination center of the brain to fire into the rational center of the brain. (This is an oversimplification but not completely inaccurate: There is a part of the brain called the cerebral cortex that serves to dampen the deeper, older emotional centers in the limbic system, such as the hypothalamus, cingulate cortex, and the hippocampus.) That doesn't tell you whether or not Santa Claus exists, and it doesn't tell you whether or not believing in Santa Claus is good. It just tells you that there are two parts of the brain that used to not interact, and now they are interacting.

Does this mean that you would use the helmet on yourself and then believe in Santa Claus?

Should you?

If you should believe in Santa Claus, does that mean that he exists?

On first blush—yes. If everyone should believe that something exists, then it exists. Take gravity. If you want to have a good understanding of how the world works, you need to appeal to the existence of gravity. Therefore, gravity exists.

As Saint Thomas used to say, though, *"sed contra,"* which means "but on the other hand." What if it were the case that, throughout the universe, every single species that figured out it was on a planet eventually developed nuclear weapons and exterminated itself? I hope this isn't true, and don't get bummed out during a Christmas book, but consider for the sake of argument that it is. Out of a trillion intelligent races, one billion figured out that they were on planets, and all one billion of them are extinct. One could argue, then, that believing we live on a planet is a very bad idea—it will lead to us all dying! But it's still true, isn't it? It seems that for us to even know it's a bad idea to think it, we need to talk about planets, which means we have to admit that it's true! So in that case, we could argue that even if it is a good idea for us to believe that Santa Claus is real, he still isn't real. It's hard to know exactly how to put this: We could argue it but we shouldn't? We should and we shouldn't? One part of us knows he's real and one part of us doesn't? But what part is thinking *that*?

Now that I think about it, does the society of the well-meaning conspirators and the duped-for-their-own-good laity even work? How do the conspirators pass on their conspiracy to their children? If the conspiracy works, doesn't it eat up its own knowledge of itself, making it impossible to perpetuate? And if it's impossible to perpetuate once it does that, and people realize that Commonus doesn't exist, won't they fall into the tragedy of the commons?* Or will they decide to double-down on Commonus worship by tearing down the farms on the commons and building big temples to him there instead? Maybe the upkeep of the big temples will be so expensive, it will dwarf the economic costs of enclosing the commons. How could you

* Maybe that's why there are historical cycles.

keep that from happening? Once you make people crazy, who knows where it will lead!

On the other hand, if the conspiracy works, what keeps the conspirators from conspiring about other things, taking advantage of their position to feather their own nests and scam women? For the system to run smoothly, there needs to be somebody with a knowledge of just how conspiratorial the conspirators can be, and just how duped the common people can be, and there is precisely nobody who can hold this knowledge. Similarly, if I know that spontaneity is an intrinsic side benefit, I can't achieve spontaneity—there can be no part of me that carries the knowledge that lets me be spontaneous. Does that mean that spontaneity is an intrinsic side benefit, but nobody can know that?

How about Santa Claus? If we have decided that the best thing is to lie to ourselves and believe in Santa Claus, do we believe in Santa Claus or don't we? It seems that we do and we don't, both as individuals and as a society. But this was exactly where we started. The path of logic has led us nowhere.

It's time to consider whether the initial move of logic was a mistake. Maybe saying and believing self-contradictory things isn't a sign that we're getting things wrong. Maybe it's a sign that we're getting things right. There is a cross-cultural tradition, as distant as a cave in old Tibet and as close as your closest New Age bookstore, whose proponents say we should embrace self-contradiction because it tells us something very deep and important, namely, that reality transcends logic. Life cannot be embraced by human thought or language, and life can be both A and not A. This alternative to logic is called mysticism, and we will now turn to what mysticism has to say about Santa.

Cue sitar.

PART 2

MYSTICISM

4

Finger, Meet Moon; Moon, Meet Finger

I grew up in an old Victorian house in an unfashionable and semi-dangerous corner of 1970s Brooklyn—no murders on my block, but in front of the candy store on the avenue, sure. My father had a storefront law office with a big sign in front that said ABOGADO, guarded by an iron gate and an electric eye because junkies had once broken up through the floor to steal his copy machine, and my mother worked at the local high school teaching biology to teenagers, many of whom just wanted to use the biology they had already mastered to have babies. My family was not good at expressing or communicating emotion—my parents went through a tragedy before I was born, and the emotions they would have expressed, had they expressed emotions, would have been too painful. So family discourse around the dinner table centered on logic,

the approach I laid out in the first section. But there was another one in the air, namely mysticism.

I learned about mysticism from hippies and my local library. Though they flourished in the United States in the 1960s,* there were still plenty of hippies around in the 1970s. They espoused a philosophy of seeing past the limitations of the rational mind through the use of drugs—mostly marijuana and psychedelics. Even though I was too young to go to hippie parties, I did run

* And have roots that go back at least to nineteenth-century German nudists and naturists, if not much further.

into hippie philosophy at the hands of my babysitter's boyfriends, who would sometimes engage me in debate.

This debate would usually take the form of the hippie challenging me to justify my commitment to rational thought. "Why are you allowing yourself to be limited by thought and language? How do you know that the way you think is right, since other people in other cultures think differently? You need to get out of your uptight rational mind and have an experience that transcends it." The hippie would then get bored with arguing with an eight-year-old and make me watch TV while he went off to have sex with my babysitter.

I thought the babysitter's hippie boyfriend had some good points, so I went to my local library (which featured a guy with Tourette's who would sit reading and burp unpredictably—just

when you thought he wasn't going to burp, he would) and borrowed a book on philosophy that featured a few pages summarizing the ancient Hindu holy book the Upanishads, which more or less said I was wrong and my babysitter's boyfriend was right. The highest or ultimate reality according to the Upanishads was the Self, because it was inside me and also the ground of the universe.

The Self "is not born, nor dies. It is not from anywhere, nor was it anyone. Unborn, everlasting, eternal, primeval, it is not slain when the body is slain." (Roebuck, Katha Upanishad 2:18)

How do we get it? Not by thinking!

When, making the mind thoroughly firm,
Free from laxity and distraction,
One reaches a state without mind,
That is the highest state. (Roebuck, Maitri Upanishad 6:34)

When I read this as a kid, it sounded great. Meaning and immortality? Those were, like, my two favorite things! As an even littler kid, I had become very upset by the fact that I was going to die. I would lie in bed at night and try to imagine not existing. Then, when I had gotten good and scared, I would realize that I hadn't successfully done it—that I was still imagining myself there experiencing my nonexistence. So I would try to imagine that that wasn't there either, and would get even more scared. Then I would realize that even that wasn't really imagining my annihilation, because there was even a person there imagining it, so I would run to my parents' room and squeeze under their bed.

Logic wasn't much help. In fact, it just made things worse, because what logical reason was there to believe I would exist after death? And what logical reason was there to believe that, given that, anything had any point? I would die, everyone would die, the universe would end. Why even get out of bed? Mysticism seemed much more promising! I got together with my friend Jonathan Blaine in junior high school before we went to work, and we sat on the mattress in his attic family room cross-legged and tried to achieve the Self. Later, I studied Buddhism, became a Buddhist monk at Wat Chulamani in Thailand, and got a mantra from an Indian guru who traveled the world hugging people. I wanted to have the experience that the Upanishads and my babysitter's hippie boyfriend were so enthusiastic about.

What does the mystic learn about reality from his mystical experience? It's hard to say, because it's impossible to say. The mystic's insight literally can't be communicated in words. But when the mystics try to put it into words, they say things that are self-contradictory. So you could say, to the extent that it can be said, the reality that the mystic comes to know contradicts itself. So as mystics, we don't try to avoid self-contradiction as we do in logic. We come to an understanding that reality is self-contradictory.

In the paradoxically titled *De Docta Ignorantia—On Learned Ignorance*—the hermetic philosopher Nicholas of Cusa says that reality is characterized by a *coincidentia oppositorum*—a coincidence of opposites. The Upanishads agree:

It moves, it does not move;
It is far and near likewise.
It is inside all this:
It is outside all this. (Roebuck, Isa Upanishad 5)

Not only is reality contradictory, but so is the process of coming to know about it:

It is thought of by the one to whom it is unthought;
The one by whom it is thought of—he does not know.
It is not understood by the understanders;
It is understood by those who do not understand. (Roebuck, Kena Upanishad 2:3)

The Upanishad seems to mean that those who understand it don't understand it, while those who don't understand it do understand it. That's hard to understand! Maybe if we don't understand it, we do! But now we think we do, so we don't! Logic tempts us to dismiss this as drivel, but who says we should trust logic? If we're interested in the data, they state that a lot of people in different epochs and cultures, when they've wanted to share their deepest insights into life, have said things that are self-contradictory. Self-contradictory but, they claim, TRUE.

In Christian theology, Dionysius the Areopagite writes,

"Unto this Darkness which is beyond Light we pray that we may come, and may attain unto vision through the loss of sight and knowledge, and that in ceasing thus to see or to know we may learn to know that which is beyond all perception and understanding." (Dionysius the Areopagite 194) This isn't just for hippies; some of the most orthodox of the orthodox Catholic theologians embraced this negative theology. (Although you could view monks as sort of like institutionalized hippies.) According to Saint Thomas Aquinas, strictly speaking, no predicates are able to describe God. He's not really "good," he doesn't really "exist"—these are just analogies to help us understand a reality that is beyond the mind.

It almost seems that, to understand this, you'd need to consult a philosopher who took a lot of drugs. That's true. The best philosophical account of mysticism comes from William James in his discussion of his experiences, not on LSD, which hadn't been invented yet, but on nitrous oxide, a gas that high school kids in my neighborhood called whip-its because it could be sucked out of whipped cream canisters. James said the mystical experience he had on nitrous was both *noetic* and *ineffable*, by which he means it causes you to *know something* but *something that can't be put into words.* What does being "put into words" mean exactly? Felix Mendelssohn, who composed *Songs without Words,* said that the thoughts he expressed in music were not expressible in words because the words were too coarse-grained. But, you might argue, maybe they can be put into words because you can use the words "the thoughts Mendelssohn was expressing in the Venetian Boat Song." James says the feeling of déjà vu is noetic but ineffable, and so is the feeling of sudden insight into something we knew before but never, until that very moment, truly felt. Neither experience can be put into words, but they cause us to know something we didn't know before we had them.

Recalling his trip on whip-its afterward, James writes:

> One conclusion was forced upon my mind at that time, and my impression of its truth has ever since remained unshaken. It is that our normal waking consciousness, rational consciousness as we call it, is but one special type of consciousness, whilst all about it, parted from it by the filmiest of screens, there lie potential forms of consciousness entirely different. We may go through life without suspecting their existence; but apply the requisite stimulus, and at a touch they are there in all their completeness, definite types of mentality which probably somewhere have their field of application and adaptation. (James 388)

So James believes that there is an ineffable experience he had when he was high and it has revealed to him that the brain is capable of more ways than one of cognizing reality.

> Looking back on my own experiences, they all converge towards a kind of insight to which I cannot help ascribing

> some metaphysical significance. The keynote of it is invariably a reconciliation. It is as if the opposites of the world, whose contradictoriness and conflict make all our difficulties and troubles, were melted into unity. Not only do they, as contrasted species, belong to one and the same genus, but one of the species, the nobler and better one, is itself the genus, and so soaks up and absorbs its opposite into itself. This is a dark saying, I know, when thus expressed in terms of common logic, but I cannot wholly escape from its authority. (James 388)

And my babysitter's hippie boyfriend would agree.

But what does that have to tell us about contradiction? Can two contradictory approaches both be true? Does life have a point and also not have a point? Does Santa Claus exist and not exist? If the mystic has learned something that he can't communicate, how are we supposed to know what it is? If it's ineffable, then how could some sentences say it and not others?

What if my brother Flippy, playwright, bank officer, father of two, who used to beat me at wrestling and chess and then stopped when I got better than him at both, is actually the world's greatest mystic? How could he tell me if he was or he wasn't? Maybe when he says, "I've just scanned some slides of our family trip to the Grand Canyon and posted them to Facebook," he is really doing the best he can to express his ineffable insight into the All? I hope the problem is not that my brother Flippy doesn't wear robes and have a long beard and a shaved head, because that's incredibly shallow, and mysticism is supposed to be deep! Maybe the problem is that "I've just scanned some slides, etc." doesn't sound mystical enough. But what does it mean to "sound mystical"? Well, you can usually find things that sound mystical written on little wooden signs or smooth stones available next to the cash register at the mailbox store, like, for example, "The Limitless is all around you, just look." If you paint "I just scanned some slides of the Grand Canyon and posted them to Facebook" on a little piece of wood and try to sell it at the mailbox store, nobody will buy it because it doesn't sound mystical enough. But if "I've just scanned some slides" doesn't mean the inexpressible thing and "The Limitless is all around you" does, and that difference can be discerned by the patrons of the mailbox store, then doesn't it follow that the mystic's inexpressible insight is at least somewhat expressible? And in this hard-hearted, fast-paced, go-go world, doesn't "somewhat expressible" just mean "expressible"? Somebody expressed it well enough to sell a little painted slogan, after all. I mean, who would even want their meaning to be "completely expressible"—wouldn't you want to leave a little inside, yet to be pressed out, for later? Flawed and inadequate as it may be, language is still one of our all-time top tools for saying things, and if we want to get a grip on what the mystic is saying, we will have to use it. But how?

5
Santa Yoga

The mystic sometimes compares his language to a finger pointing at the moon. So even though the advocates of mystical traditions sometimes claim they disagree with each other, they might be wrong. So Buddhists will sometimes tell you that they differ from Hindus because Hindus believe that all that exists is the Self while the Buddha taught that the self is an illusion. However, it's not clear to me that the Buddhists have a right to tell you this, because what the Hindus have to say about the Self is that there is nothing you can say about it. It's "*neti, neti*"—which has nothing to do with the nasal irrigation pot but is Sanskrit for "not this, not this," as in "Is it the mind?" "Is it the body?" "Is it time and space?"—answer: not this, not this. So maybe the reality that the Hindus believe you can't talk about, but nonetheless call the Self, is the same reality that the Buddhists believe you can't talk about and call the

Non-Self. In fact, forget the *maybe*. If that reality is all that exists, it has to be the case. The two realities have to be one reality because there are no two realities. The Hindus are pointing at the moon with their thumb, and the Buddhists with their pinky finger. And the Tantrics with something else.

The idea that there could be some experiences that are ineffable brought me into conflict with the biggest philosopher at Berkeley when I was there: Donald Davidson. At that time, the mid-1990s, Davidson was considered so important that students would make pilgrimage from places as far-flung as Taibei and Hamburg to write monographs exploring one tiny wrinkle of his carunculated cerebral cortex. Davidson had a philosophy that stressed interpretive charity but, ironically, was angry at everybody who disagreed with him, and interpreted his philosophical rivals as a pack of fools. This combination of methodological charity and real-life contempt made him a formidable opponent. Following his mentor, Professor Willard Van Orman Quine, Davidson argued that it was impossible for anybody to have a radically different concept of reality, and that the concept "concept of reality" didn't even make sense, because in order for us to understand somebody as having a language at all, we have to interpret what he says, and to interpret what he says, we have to believe that it's about reality: objects, water, salt, pepper, and such. I found this quite upsetting because at the time it seemed to me that it made mysticism impossible. And that was by design: Davidson and Quine scorned discussion of the "inner light" and Alfred North Whitehead, Russell's more mystical writing partner. (Quine said, after listening to a speech of Whitehead's, "Alfred North Whitehead, Mary Baker Eddy, Jesus H. Christ.")

In his paper "On the Very Idea of a Conceptual Scheme," Davidson in particular devastated the thesis and reputation of Benjamin Whorf. Whorf was an engineer and amateur linguist who studied the Hopi and concluded that they didn't have a Western notion of time, and Whorf's studies of the Hopi way of describing time gave impetus to the whole one-sided love affair between hippies and Hopis. The Quine/Davidson thesis was essentially that we would have to listen to the Hopi language and assign some meaning to its tense system, and if our Hopi informant used one word to talk about something that happened yesterday and a different word to talk about something that would happen tomorrow, we would have to find that they had a concept of time.

I asked my cousin Harry the hippie and linguist and hoped he would take my side against Davidson, but he didn't! He though Davidson was right! Nevertheless, even without cousin Harry's moral support, I struggled against Davidson's philosophy all through graduate school, and in a fit of hubris, I had Davidson on my oral committee. He passed me, albeit grumpily, possibly because he was hungry. I felt I had to take on Davidson because my hopes of mystical insight, and thus happiness, and thus life after death, were riding on it. If Davidson was correct, I would never find an American Indian shaman, or Tibetan lama, or mystical woman with long black hair who has an insight into reality that goes beyond what I've experienced growing up in my family's weird, sad house in Brooklyn. Which would be very disappointing!

I really want to believe that mystics are not just flimflam men, or confused, fuzzy-minded people who would have failed Introduction to the Philosophy of Language when Davidson taught it at Queens College. Is there any way to get a grip on their message? The Buddha sometimes talks as if he could say what he knows, but he doesn't want to, because he thinks it's not important. In the Cula-Malunkyovada, or Poison Arrow Sutta, he says that he teaches the arising of suffering and the cessation of suffering, and compares his followers who ask questions about reality to people who, upon being shot by a poison arrow, want to know the archer's hair color before getting to work extracting the poison. It's sort of a weird point. First of all, even if life is suffering, we're not in *that* much of a rush, especially if time is an illusion. The Buddha ought to be able to answer our questions about the universe for like ten minutes, and then tell us how to be free of suffering. And secondly, suffering is in a sense a false belief in the self as a free-standing entity, so you would think that extracting the poison would be pointing out that this belief is false and replacing it with a true one. But although the Buddha does tell us that our naive belief in our self is false and getting us in trouble, he never gives us a more correct belief to have about ourselves. What he does is go down all the possible other ways we can think of to think about the issue and inform us that they are all wrong. The Pali Canon, which is the best record of what the Buddha actually said, records this conversation between the Buddha and a follower on the question of what happens to a monk once he is released from the illusion of belief in a free-standing ego:

> [Aggivessana Vacchagotta:] "When a bhikkhu's mind is liberated thus, Master Gotama, where does he reappear [after death]?"
>
> [The Buddha:] "The term 'reappears' does not apply, Vaccha."
>
> "Then he does not reappear, Master Gotama?"
>
> "The term 'does not reappear' does not apply, Vaccha."
>
> ". . . both reappears & does not reappear, Master Gotama?"
>
> ". . . does not apply, Vaccha."
>
> ". . . neither reappears nor does not reappear, Master Gotama?"
>
> ". . . does not apply, Vaccha."
>
> "When Master Gotama is asked these four questions, he replies: . . . 'reappears' does not apply . . . 'does not reappear' does not apply . . . 'both reappears and does not reappear' does not apply . . . 'neither reappears nor does not reappear' does not apply. . . . Here I have fallen into bewilderment, Master Gotama, here I have fallen into confusion, and the measure

of confidence I had gained through previous conversation with Master Gotama has disappeared." (Nanamoli/Bodhi 592–93)

Poor Vaccha! Why doesn't the Buddha just come out and say what he has to say? Is he being coy? Is he employing "bullshit"?—a term the philosopher Harry Frankfurt defines as spinning words with no commitment to their truth or falsity in order to advance a personal agenda. That's hard to believe too, because according to the legend, the Buddha had every conceivable personal advantage—wealth, power, status, a harem of women, a horoscope that gave him a straight shot at becoming emperor of India—and gave them up because he realized they wouldn't bring him lasting happiness. Now, we could think the legend is just PR, but why not give the hundreds of millions of people who have found contentment for centuries from the Buddha's teachings a little credit? If we, for the sake of argument, consider that they are not all being conned, how would that work? How can what the Buddha says be at the same time completely true and completely slippery?

Maybe the Buddha stands in the same relationship to a regular person as a regular person does to Ebenezer Scrooge. Scrooge is somebody who thinks only about money. He has no close friendships, doesn't love anyone, and does nothing but make money and watch his bank accounts go up. He has a very limited emotional life and no imaginative life or spiritual life at all.

Imagine we are trying to convey to him our feelings about our children. Scrooge thinks that everything is determined by its cost. We say, "My love for my child is not something I would ever sell." Scrooge says, "So it's worthless!" We say, "No, it's not worthless. It's of a higher value than anything!" Scrooge says, "Aha. It's super expensive! Just like a sports car but much higher." We say, "No, it's not like that. It's neither of no value nor of high value." Scrooge says, "Okay, so it's *both* of high value and of no value." "No!" "It's *neither* of high value nor of no value?"

No, Scrooge! Wrong! Scrooge is just in a constricted, neurotic weird state of consciousness—maybe he can use some nitrous oxide or LSD (the hippies certainly thought so!). We are in a higher state of consciousness than he is, and therefore our language is going to be hard for him to wrap his head around. We are better, and we use language and thought better, and we are able to think things that he can't (yet) think. What he needs to do is expand his consciousness, to get into a different

headspace or, in the terminology of the Hasidim, to move from contracted consciousness to expanded consciousness.

If we want to help Scrooge, we can offer a therapy to him to get out of his money-consciousness. We can tell him to go for a walk and not think about money so much and just appreciate the beauty in his life. He may try that and it could be quite a struggle—as he sees a beautiful ray of sunshine, he thinks first, "Wow! Awesome!" And then, "Hey, I can put in solar panels here and get a tax write-off!" We can give him poems and music written by people who are not obsessed with money, and they can stir him emotionally. However, he might misunderstand these and just decide that he will collect poems and music and philosophy written by people who don't care about money—and even get a really valuable collection someday that he can sell on eBay for millions! There's a sense in which Scrooge, the money monomaniac, can be led by baby steps to realize there is more to life than money, but there's also a sense in which none of these baby steps will ever be able to do it—he has to make a leap into a new form of consciousness, a new form of being. If he succeeds in doing so, then Scrooge will say his life is better, he is looking at a much more beautiful universe, and his old way of looking at things will seem dark and partial. Before he had his enlightenment experience, Scrooge saw us doing and saying things that didn't make sense. Now they make sense to him, but that doesn't mean he can explain them in terms that would make sense to his old self.

On this view, what the Buddhist is offering is real insight, but it's real insight that our normal mistaken patterns of mind will always misunderstand. So if, hypothetically, Quine and Davidson had minds warped with anger from early childhood, and a neurotic need to control life and people with their plus-size craniums, then they would not be able to interpret what the mystic had to say. To those two, the words of the mystic would seem empty or nonsensical, but that would be a problem with Quine and Davidson, not with the mystics. You might say that certain insights are ineffable relative to some people but not to others, and the problem is that those people who find what the Buddha has to say hardest to hear are the ones who most need to hear it.

If that's true, it makes sense that there was a continual push and pull in the history of Buddhism, between the attempt to make Buddhist doctrine clear enough to be useful to people, and the worry that to make it "useful" is to give up everything that is valuable about it. So, in the beginning, the Buddha taught that there was such a thing as "nibbana"*—the cessation of ego-driven modes of thought. But later on a movement arose, the Mahayana, which maintained that by making out nibbana to be something different from regular life, their predecessors, the Hinayana,† were playing into a very non-Buddhist

---

* The Buddha probably called the extinction of suffering "nibbana" rather than "nirvana." "Nibbana" is a word in Pali, which was a spoken language. "Nirvana" is in Sanskrit, an artificial language created for literature and philosophy.

† "Hinayana," "small vehicle," is a derogatory term used by followers of Mahayana, "great vehicle," Buddhism. Hinayana Buddhists don't call themselves Hinayana Buddhists; they call themselves Theravadins—those who follow

attitude—namely that there's something out there that's *great*, we don't have it, and we need to do everything we can to get it. Just as we find ourselves having to use different modes of expression in order to bring Scrooge to an understanding that love is not just something that's worth a whole lot of money but belongs in a different category of being, so the Buddhist has to convince us that what he is talking about is not just another thing to achieve.

When the Hinayana monks expressed their central insight metaphysically, they said that all form is empty. You think there's something out there you can hold on to and really sink your teeth into, but it's actually an illusion. There's nothing worth getting and nothing worth being. But the Mahayana Buddhists felt that this approach missed the point, and expressed the following in a dialogue between the multi-headed deity Avalokiteshvara and an old-school Hinayana monk named Shariputra. Here's the money quote from the Heart Sutra, so called because it is the heart of the Perfection of Wisdom sutras, which have more and less concise forms:*

---

the way of the elders. Mahayana Buddhists say Hinayana Buddhists have a "small vehicle" because they are interested in Nirvana only for themselves, and not in saving all sentient beings, but since neither Buddhist believes the self is real, it's not clear to me how seriously to take this distinction.

* Their most concise form is the perfection of wisdom expressed in a single syllable: Ah!

Here, O S[h]ariputra, form is emptiness and the very emptiness is form;
emptiness does not differ from form,
form does not differ from emptiness;
whatever is form, that is emptiness,
whatever is emptiness, that is form,
the same is true of feelings, perceptions, impulses and consciousness. (Conze 81)

Avalokiteshvara is saying to Shariputra: "You think the Buddha taught you a different way of life, one that is free from selfish craving, and you desperately want that way of life. But thinking that way itself is a form of selfish craving. All you've done is transformed your lust for cigarettes, whiskey, and wild, wild women into a lust for enlightenment."

This point is put in more philosophical, dialectical form by a school of Mahayana Buddhist thinkers known as the Madhyamakas. The Madhyamaka school was founded by the philosopher, alchemist, and yogi Nagarjuna, who supposedly learned his philosophy under the ocean from magical snakes called nagas.

(It's wildly instructive that, while in the Garden of Eden story, the snake is a source of insight that separates us from reality, in the Nagarjuna myth the snakes are the good guys.) The magical snakes start by showing us how everything that we have to deal with is actually empty of any independent existence.

That includes us and also anything we are interested in getting or avoiding. It is sunya—empty of its own nature. It is what it is only because of its relationship to its context: A flower is a flower only because it becomes a fruit, a fruit is a fruit only because it is full of seeds, seeds are seeds only because they grow into plants, plants are plants only because they have flowers. It is impossible to say anything true about Ultimate Reality. However, the Madhyamakas distinguish between two truths—conventional truth, also known as the truth of the marketplace, and ultimate truth. So let's ask the question "Is emptiness itself empty?" Is the statement "Nothing has real existence" ultimately true, or is it just conventionally true?

Write down your answer before you read on.

If you guessed that the truth of ultimate existence was ultimately true, according to the Madhyamakas, you would be wrong. Because everything is empty, emptiness is empty. Because language cannot encapsulate ultimate reality, even the statement "Language cannot encapsulate ultimate reality" itself cannot encapsulate ultimate reality. Even the distinction between ultimate and conventional truth is conventional, not ultimate. If Wittgenstein offered us a ladder, the philosophers of Madhyamakas offer us a circular chute that takes us back where we started.

A similar paradox comes up in Taoism. In the Tao Te Ching, the foundational text of Taoism, Lao Tzu states that "Those who know do not talk and those who talk do not know." Consequently, Lao Tzu didn't know—and he was the founder of

Taoism. The Chinese philosopher Tu Wei-ming once told me that Confucius was the best Taoist because he taught about how people could get along together in society and, as following Lao Tzu's dictum, never mentioned anything about Taoism. Of course Tu Wei-ming was a Confucianist.

How does this apply to Santa Claus?

One approach would be to say that Santa Claus is just as good as anything else we believe in—America, the self, love, death, science. He doesn't really exist, but neither do they. And if we go through the Madhyamaka dialectic, we will end up believing in him—sort of. We won't be attached to him, but we can use him as we see fit. Our attitude toward Santa will be something like the attitude taken by Tibetans who practice deity yoga. They first conceive that everything is empty of any real substance and then imagine themselves to be Tantric deities: colorful, well-dressed, sexy beings with multiple heads and millions of arms for saving people.

So we might, for example, try to meditate on Santa Claus, closing our eyes and trying to have an experience of him.

We could chant the names of Santa Claus over and over again—"Santa, Santa, Santa, Santa, Kringle, Kringle, Kringle, Kringle"—a practice called japa yoga.

We could go on retreats to Santa temples full of statues of Santa, and listen to lectures about life on the North Pole.

And we wouldn't have to worry about the fact that he both exists and doesn't exist, because paradox is a real feature of reality. To paraphrase Nagarjuna:

It's not true that Santa Claus exists.

It's not true that Santa Claus doesn't exist.

It's not true that Santa Claus neither exists nor doesn't exist.

It's not true that Santa Claus both exists and doesn't exist.

And that would be that! The dewdrop slips into the shining sea.

6

I Do and I Don't

What kind of uptight jerk would have a problem with the dewdrop slipping into the shining sea?

Well, let's start by looking at James's own conclusions regarding mysticism. He argues that mystics tend to be optimistic and pantheistic. They feel that the universe is a good thing, they're happy about it and believe that everything is in some sense a manifestation of an underlying, ineffable, really quite amazing and terrific reality. He then asks the question "Does that matter? Does the fact that the mystics experience it like that *prove* that reality is so amazing and terrific?" After all, people who shoot up heroin have an insight that everything's chill and there's nothing to worry about, but they're wrong, and people who get worked up at the political rallies of whatever politician you disagree with have a guided epiphany into what's wrong with this country, but they're terribly, terribly misguided.

James argues that mystical states are and should be authoritative for the person having them, but not for the rest of us. The rest of us have to test whether the regions of reality the mystic claims to have access to are real, and whether the mystic should be emulated. He admits, for example, that there are people whose mysticism is diabolical and makes them unhappy: schizophrenics. (James, 423)

So mysticism has authority for the individual but not for other people. This formulation sounds like a good compromise—it's democratic, it's respectful, it's polite. In fact, it papers over a gigantic problem. Because the one thing the mystic seems to be trying to say most emphatically is that the boundaries between people are not real. So if James is correct that something that has authority for the mystic doesn't have authority for the rest of us, then doesn't it follow that the mystic is wrong? How could there be an experience that has authority for one of us and not for all of us, if all of us are one?

James's advocacy of mystical humility also seems not to reflect the mystic's attitude toward his own insight. Yes, there are an undisclosed number of silent mystics, but the mystics whose writings have come down to us talk *a lot*. They are super chatty about the need for silence, and really want everyone else to understand how important their ideas about selflessness are. After all, they claim that they have achieved an insight into the deepest nature of reality, and whether that comes with a booming voice or a coy chuckle, that's a big claim. Even huge blowhards like Donald Rumsfeld or Donald Trump would be embarrassed to say that, and they experience a joy at the sound of their own voices that the rest of us can reach only in the most refined regions of contemplative prayer. Even if the mystics don't say, "I'm a god, worship me" (although some do), their followers often say, "You're a god, let us worship you!" and "My guru is a god, you should worship him!" and the mystics generally don't work that hard to stop them.* The followers of mystics say we are in the same relationship to the mystical guru as the insane person is to the sane person, or the animal is to the human. If they don't think they're literally magic, and they often do, they believe their gurus are exemplars of holiness, energy, love, perseverance, and so on.

In point of fact, embracing self-contradiction is no sign of being a good person. You can in fact, just as we feared, be either a hustler trying to rip people off or a lunatic. As George Orwell illustrated in *1984*, a certain kind of paradoxical mysticism is embraced by totalitarian regimes who say, "War Is Peace" and "Freedom Is Slavery." Even today, if you want to read something that really doesn't make any sense, look up the North Korean philosophy of juche. Recent books on scientology reveal how L. Ron Hubbard brutalized and confused his cultists with illogic, throwing them off ships and making them traipse around

---

* Some people say they can't convince their followers that they're not gods, but really? If I had to convince people I wasn't a god, I bet I could. I could eat a lot of pizza and throw up on them, for example, or just confess to specific, embarassingly foolish decisions.

the world looking for buried treasure from previous lifetimes, and George Gurdjieff was awful to his followers as well.

Buddhists often make the claim that when we realize that the self is an illusion, it makes us less selfish. Nagarjuna wrote *sunyata karuna garbah*—"Emptiness is the womb of compassion." That is, the realization that all selves are empty and interrelated should make us loving and compassionate. Robert Thurman, the first American ordained as a Tibetan Buddhist monk (and a former teacher of mine), writes in his introduction to the Vimalakirti Sutra, "[Buddhist teachers] recommend their full cultivation of great love and great compassion while maintaining total awareness of the total absence of any such thing as a living being, a suffering being, a being in bondage. In short, they show the way to the full nonduality of wisdom and great compassion." (Vimalakirti 6)

It doesn't have to, though, because as we saw in our discussion of Madhyamaka, the original conventions get reappropriated. Buddhists will say that when they get reappropriated, they get reappropriated in a calmer, less crazy way. But why? Shouldn't they be reappropriated in a way that's 100.000000 percent as uptight and insane as they were to begin with? If the path is truly a circle, shouldn't it take us back to exactly where we started?

Thurman continues: "Wisdom doesn't allow us to settle for our habitual involvement with sensory objects (as just being 'given' 'out there') and causes us to learn and practice probing beneath the surface of apparent 'reality' to gain direct awareness of the ultimate reality of all things. At the same time, great compassion does not allow us to set up any hypostatized 'ultimate reality,' immerse ourselves in any sort of quietistic trance or accept any sort of illogical escapism from relativity, but imperatively compels us to act selflessly." (Vimalakirti 3–4)

Vimalakirti's argument as presented by Thurman is that understanding emptiness and then the emptiness of emptiness will give us a freedom from obsessive thought and obsessive action, and that will make us compassionate. But consider the following. If I hear somebody screaming, "Help! Help!" from the apartment upstairs, if I'm a good, caring person, I feel impelled to go help. But if I know it's just an MP3 player playing a .wav file of someone saying, "Help! Help!" and there's no person there, then I'm not impelled to do anything. Shouldn't the Buddhist consistently feel the same way: that it makes just as much (or as little) sense to not help as to help?

This isn't just theoretical. If you went over to a Tantric Buddhist yogi's for supper, take a close look at what's on the bill of fare before digging in. As the Yogaratnamala (Necklace of Jewels of Yoga) explains, "Food and drink should be had as it comes and not be rejected by thinking in terms of what is acceptable and what is prohibited. One should not perform the rituals of bathing and cleansing or avoid vulgar behavior. . . . He should eat all kinds of meat. . . . He enjoys with all kinds of women having a mind free of all trepidation. . . . He must eat the Five Nectars, drink liquor made from molasses, eat the poisonous Neem, and drink placental fluids. He must eat foods which are sour, sweet, bitter, hot, salty, astringent, rotten, fresh

and bloody liquids along with semen. By means of the awareness of non-dual knowledge there exists nothing inedible. Obtaining menstrual blood, he must place it in a skull cup and mixing it with phlegm and mucus, the holder of the Vow must drink it." (Farrow/Menon 198)

It's not just about diet either. "You should kill living beings, speak lies, take what is not given and service others' wives." (Farrow/Menon 192)

Now, if a Tantric Buddhist wants to get over his dualistic consciousness by drinking menstrual blood and phlegm and mucus out of a skull, who am I to say no, as long as he got permission from the original owners of the skull, the mucus, and the menstrual blood. I might even give him the phlegm myself, if I'm feeling generous. As long as nobody gets too sick from the poisonous Neem, everybody, it's going to be a party we'll all remember for a long time. But I'm warning you, if he wants to start murdering people, and I'm one of the people he wants to murder, I'm going to criticize his philosophy.

Some of the nicer Tantric Buddhists, for example, the Dalai Lama, interpret all this talk of murder and fornication as a metaphor. But not everyone. The Tantric Buddhist leader Osel Tendzin had unprotected sex after he knew he had contracted HIV, spreading the disease to male and female followers.* This was condoned by some in his organization because, as a higher spiritual being who had transcended duality, the dharma regent could do this in order to open his students' minds. Thanks, Osel!

Another problem with mystics is that when they are not being authoritarian, they can be fatuous. When a student asked the late Columbia philosopher Sidney Morgenbesser if he agreed with Mao Zedong that something could be both A and not A at the same time, Sidney said, "I do and I don't." Which is to say that while logic and inquiry have a way of helping us approach reality and improve on our beliefs, a standpoint that is pro contradiction and pro mysticism seems to not move the ball forward at all. If you want to know whether there are wild marsupials in North America, it doesn't help to be told there are and there aren't. It does help if somebody says there are. You can then say, "Prove it!" and they can show you an opossum. Morgenbesser's point is a restating of Aristotle's, who says it is as useless to argue with someone who does not accept the law

---

* Butler, Katy. "Encountering the Shadow in Buddhist America." *Common Boundary Magazine* (May/June 1990). http://www.katybutler.com/publications/commonboundary/index_files/commbound_shadowbuddhistusa_new.htm.

of noncontradiction as it is to argue with a vegetable. You wouldn't want to argue with a vegetable and you wouldn't want to marry someone who at the altar said, "I do and I don't."

It makes sense that mystics can be both fatuous and authoritarian because one of the chief weapons the mind can use in combating unjust authority is to point out a contradiction. And if mystics respond by saying, well, that's just the way it is—"Life is contradictory"—echoing our parents when they've gotten tired of justifying a punishment and say, "Life is unfair"—then they checkmate one of the mind's best strategies for upsetting an unfair or superstitious status quo. It's worth noticing that India, the culture that gave us the Upanishads also gave us Shitala, the smallpox goddess. When Indian children died of smallpox, their parents flattered the goddess and said nothing could be done. On the other side of the Eurasian landmass, Edward Jenner noticed that cow-maids who had had cowpox never got smallpox, invented vaccination, and smallpox was eradicated. Point for rationalism!

Sometimes the distinction between self-effacing Zen simplicity and fatuousness can be razor thin. As we saw, the first formulation of Buddhism, the so-called Hinayana, said that the ego is an illusion and suggested we live as monks, free of property, family, sex, and dinner,* in order to free us of the illusion of the self. After Hinayana, the Mahayana argued that Buddhist

* Dinner is a big one. According to the Vinaya, the code of monastic discipline for Buddhist monks, a monk may not eat after noon. The Thai sangha is split

monks were building up the very psychic structures they should be tearing down, by telling people there is a great thing called freedom from suffering, and you should try to get it. The Mahayana Buddhist argues that when they do this, the Hinayana Buddhists just confirm people in their greed but give them something new to be greedy about: greedlessness. Nevertheless, the Mahayana Buddhist still looks Buddhist. He teaches people how to follow the Buddha's teachings, dresses in robes, and meditates, although sometimes he will meditate on the koan "If I'm already enlightened, why am I meditating?" On the next step along this road are some teachers* who say not only should you not try to reach nibbana, but also you shouldn't even believe in Buddhism. These Buddhist teachers are simply applying the same critique to Mahayana that Mahayana applied to Hinayana, warning us that we shouldn't be unhealthily attached to anything, including the view that we shouldn't be attached to anything. At this point, though, you may wonder why they should even get the mic if, by their own account, they don't even have any views that are worth listening to. It reminds me of a friend of mine who was very interested in a French philosophy called deconstruction. He advertised to me as one of deconstruction's selling points that deconstruction deconstructs itself. I couldn't help responding, if deconstruction deconstructs

on the issue of whether it is okay to have drinkable yogurt: The more lenient Mahanikaya say it is, but the stricter Thammayut branch say no.

* For example, Toni Packer.

itself, why bother reading its long, boring books? Why not go for a jog instead, or reread one of Patrick O'Brian's tremendous tales of the sea?

Consider for a moment the case of Krishnamurti and the Theosophical Society. The Theosophical Society was founded by Madame Blavatsky, a Russian charlatan who claimed to be channeling the teachings of magical adepts in Tibet, including one Koot Hoomi. "Charlatan" may seem a little harsh for a gentle tome of yuletide cheer, but suffice it to say real Tibetan names sound nothing like Koot Hoomi, Blavatsky's descriptions of Tibet are nothing like what we know was really going on in Tibet at the time, and her doctrine, put forward in *Isis Unveiled,* is a lot of science fiction stuff about sunken continents and lost races of men with magic powers. Blavatsky's successor as leader of theosophy, C. W. Leadbeater, identified a gorgeous young Indian man named Krishnamurti as an incarnation of Maitreya and a spiritual teacher of the world. After a few years of this, Krishnamurti renounced his messiahship and told the members of the Theosophical Society that the belief in messiahs and spiritual societies and adepts with special wisdom was a trap.

Good for Krishnamurti! I think it shows admirable fellow-feeling and public spiritedness to stop a bunch of people who want to say you're God—to ignore the maxim of *Ghostbusters* that if somebody says you're the incarnation of Maitreya, agree with them.

However, even though Krishnamurti denied he was the incarnation of Maitreya, he continued to teach, saying things like this:

> That state of mind which is no longer capable of striving is the true religious mind, and in that state of mind you may come upon this thing called truth or reality or bliss or God or beauty or love. This thing cannot be invited. Please understand that very simple fact. It cannot be invited, it cannot be sought after, because the mind is too silly, too small, your emotions are too shoddy, your way of life too confused for that enormity, that immense something, to be invited into your little house, your little corner of living which has been trampled and spat upon. You cannot invite it. To invite it you must know it and you cannot know it. It doesn't matter who says it, the moment he says, "I know," he does not know. The moment you say you have found it, you have not found it. If you say you have experienced it, you have never experienced it. Those are all ways of exploiting another man—your friend or your enemy. (Krishnamurti 122)

It's hard not to view this as ironic, like Socrates's claim that he knows nothing. After all, why is he insulting me, calling my emotions shoddy and so on, if he doesn't know a thing or two better than I? Sometimes he even chastises his audience for being too violent and lectures them on why they are:

> If you want to stop violence, if you want to stop wars, how much vitality, how much of yourself, do you give to it? Isn't it important to you that your children are killed, that your sons go into the army where they are bullied and

> butchered? Don't you care? My God, if that doesn't interest you, what does? Guarding your money? Having a good time? Taking drugs? Don't you see that this violence in yourself is destroying your children? Or do you see it only as some abstraction? (54)

So even though he denies being the messiah or Maitreya, Krishnamurti does think we're callous, insufficiently concerned about world peace, hypocritical, lazy, and asleep, and even though he also says that anybody who thinks he knows anything is wrong is doing violence to his listeners, he is talking as if he knows something we don't, namely the right attitude we should take toward violence.

It's as if, although he no longer claims to have the content of a religious leader, he still has the forms, and the forms supply the ghost of the content: There is a secret knowledge; it's really important to have it; and he has it and we don't. Even though he protests that such a position is wrong. If he really believed he had nothing to say, he should have at least shut up or at least not been so preachy. He would not claim, I think, to know something as important as how to end violence. Since he didn't shut up, it's hard to believe his claim that he has nothing to say and is not putting himself forward as a guru.

Consider this passage:

> Through complete negation alone, which is the highest form of passion, that thing which is love, comes into being.

> Like humility you cannot cultivate love. Humility comes into being when there is a total ending of conceit—then you will never know what it is to be humble. A man who knows what it is to have humility is a vain man. In the same way when you give your mind and your heart, your nerves, your eyes, your whole being to find out the way of life, to see what actually is and go beyond it, and deny completely, totally, the life you live now—in that very denial of the ugly, the brutal, the other comes into being. And you will never know it either. A man who knows that he is silent, who knows that he loves, does not know what love is or what silence is. (Krishnamurti 124)

A paradoxical utterance from Krishnamurti. I find two voices within me that respond as follows:

> **Voice One:** Well put, Krishnamurti! Love, humility, and silence are real and they're great and it's trying to grab hold of them with the mind that gets me in trouble.
>
> **Voice Two:** Nonsense. Either "love" and "humility" and "silence" mean something, and we have some way of recognizing things they apply to and things they don't, or these are just empty words. Krishnamurti is either saying something, in which case he is wrong, or he is not saying anything, in which case he is not saying anything.

At this point I feel like my dog when he spots his tail and decides to give chase: Although I've run and run fast, I haven't caught a thing. We started with two voices in conflict about Santa Claus.

Voice One: There is Santa Claus.

Voice Two: There ain't no Santa Claus!

And this pattern has come back again on a higher level.

Voice One-B: Get yourself out of the fight between Voice One and Voice Two by figuring out which one is right, by using logic.

Voice Two-B: Don't worry about the fight between Voice One and Voice Two—embrace the fact that reality is contradictory. Be a mystic.

And needless to say, this argument can continue as far as you want to take it. Like my dog, we can chase our tails for as many revolutions as we have the stomach for, but ultimately we will flop down exactly where we began. Thus:

Voice One-C: Use logic to ajudicate the claims between Voice One-B and Voice Two-B.

Voice Two-C: Stick with your mysticism, bro.

But what does being a mystic really get us? We want, I think, some way to coexist fruitfully with people who believe in Santa (or that life has a point) and with the parts of ourselves that believe in these things. And it seems that when the mystic says concepts don't reach reality, he is putting an end to ideological battles. Everyone is equally right and everyone is equally wrong. Santa Claus, God, human equality, and justice all exist. But just as the Romans were justly scolded by Tacitus for making a wasteland and calling it peace, so the mystic brings ideological war to an end at the cost of a general devastation. If everything exists, then we are in trouble because alongside God, human rights, and Santa, the following things exist as well:

Manta Claus—just like Santa but a manta ray
Mantis Claus—just like Santa but with a praying mantis head

Human Rights for Everybody but Eric Kaplan, who should get smacked in the head
Mylanta Claus—just like Santa but instead of toys, he brings Mylanta to good little boys and girls with acid reflux
God who damns everybody whose name is Joshua
God who saves everybody whose name is Joshua
Human Rights for Red-Haired People, and everybody else has to be their slaves
Hantavirus Claus who comes on Christmas Eve bearing sacks of infectious rodent excrement
And so on.

In other words, if everything exists, then the question of whether or not something exists doesn't matter.

We need to somehow approach the contradictions of our lives without denying them, but without embracing them either. They should be spurs to intellectual and emotional growth. While the mystic says it's perfectly fine that things exist and don't exist, and the logician says it's nonsense that things exist and don't exist, we need a third option. But what could that possibly be?

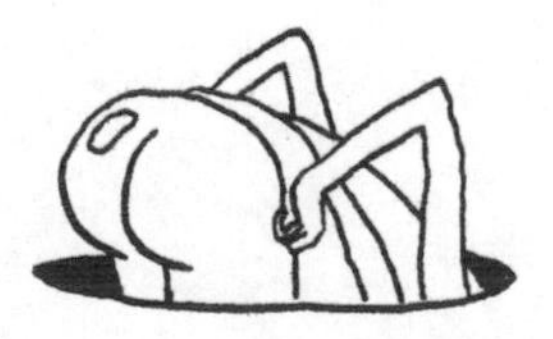

PART 3

COMEDY

7
The Jellies of Santa and Silverman

When I told my sister-in-law that I was working on this book on Santa, and that I wasn't landing firmly in the "Santa exists" or "Santa doesn't exist" camps, she said, "Sure, I get that. You're ambivalent." And I thought, no, that's sort of the opposite of what I'm saying. I'm trying to come up with a way to engage actively with two opposing realities, while being ambivalent is just dithering between attraction and fear. That made me realize that even if we accept that successfully engaging with reality has something to do with being able to embrace contradiction, it doesn't end our investigation, because there are different ways of engaging with the two horns of a contradiction, some better and some worse. We don't want to engage with the two horns of a paradox by being hypnotized by them or shutting down in a panic. We don't want to become a split personality who sometimes is fixated on one side of the paradox and sometimes "flips" and becomes zealously attracted to the other, and when we like one side, persecute people who like the other, with the savagery of self-mutilation. We don't want to just withdraw and we don't want to sink into confusion.

So we are looking for an approach to Santa Claus that is also going to be an approach to contradiction in general and one that brings together what is best in logic and mysticism. Luckily, Santa himself gives us a big hint. After all, if we know anything about Santa at all, we know the following:

His eyes—how they twinkled! his dimples how merry!
His cheeks were like roses, his nose like a cherry!
His droll little mouth was drawn up like a bow. . . .
He had a broad face and a little round belly,
that shook when he laughed,
like a bowlful of jelly. (Moore, C. C.)

Droll. Merry. Powerful laughter causing stomach to shake like jelly. Santa is letting us know that the approach to paradox we are looking for is comedy.

Although Democritus was called the laughing philosopher, the Western rationalist tradition has been pretty bad on comedy. The philosopher Thomas Hobbes says we laugh when we suddenly feel superior to someone else.

"Sudden glory is the passion which makes those grimaces called laughter. It is caused in people either by some sudden act

of their own that pleases them, or when comparing themselves to others, people suddenly applaud themselves because of the apprehension of some deformed thing in the other." (Hobbes 34) In other words, we watch a nobleman ride by on a horse and feel he is better than us. He falls off the horse and we laugh, because we suddenly feel better than him.

This is a pretty unfunny caricature of the comedic spirit, and unsurprising coming from a thinker as obsessed with domination as Hobbes. Nevertheless, he grasped a key fact. Laughter approaches contradiction by using time. *First* the horseman seems better than us, and *then* he seems worse. Our viewpoint changes suddenly. Logic and mysticism in their own way invite us to step outside of time, but comedy walks us from a state of tension to the release of laughter. In so doing, I would argue, it heals an internal split without denying either side of the split. It is a third atittude toward contradiction to place alongside logic and mysticism.

Let's take an example. Suppose Brendan is an antibullying educator with a masters in psychology who works at the district and has written numerous articles on the psychology of bullying. When he's not doing research or speaking on public television, he travels to elementary school classrooms and puts on puppet shows with his character Jimmy the Bully. Suppose we have all been called into a special assembly. Our principal introduces Brendan and he takes the stage. Brendan says he would like us to meet someone and reaches into his bag for Jimmy. As he does so, we hear the unmistakable sound of Brendan involuntarily relieving himself in his pants. Although for Brendan this would definitely be a very bad day, for most of the audience of ten-year-olds, it will be the funniest thing that they ever saw, and a memory that continues to bring joy well into deep old age. Why?

We laugh at jokes that touch on things about which we have powerful and conflicting feelings. Laughter is a release of tension, although when we relieve ourselves by laughing, it's different from when we relieve ourselves by relieving ourselves. For these kids, abstaining from pants defecation is a recent, hard-won achievement. When your colon is full, it is fun and enjoyable to relieve it, but you're not supposed to, especially in class. Authority figures—teachers and parents—are supposed, if they are masters at anything, to be the masters at the voluntary control of defecation.* Brendan defecating in his pants when he's not supposed to is funny because it acknowledges two con-

* Etymology Corner. "Defecation" is Latin for losing feces, so "fecation," the gaining of feces, must mean eating, and best not to ask the Internet for images of what "refecation" means!

flicting realities: the reality that it is fun to crap your pants and the reality that you shouldn't.

I think it's easy to gloss over this as a genuine contradiction because for many of us reading this, our toilet training is lost in the mists of time. (If for you it's a recent victory, congratulations!) But if we think about it, do we have a considered philosophy of feces? Some people find defecation a sexy component of their love play—possibly because it's a thrill to do the forbidden, maybe because they like that they can be a baby and be loved and accepted whatever they do. (We can't exclude the possibility that they like it because *it's just great.*) Once we have kids, we spend a lot of time cleaning the feces from their diapers, and if we have older parents, we may be involved in cleaning their diapers as well. It's embarrassing to begin with, but why, really? Suppose we really loved Marilyn Monroe. Would we want one of her turds preserved in gold or lucite? It would be an intimate memento.

What is the correct philosophical attitude toward this suggestion? Some mystics might welcome it. As we saw, some of them would whip out a knife and fork and dig in, in order to show that they have transcended categories of sacred and profane, yucky and delicious. I would not like to do that, and I would not like to kiss them immediately after they have performed such a demonstration of category transcendence. A logical argument as to why Marilyn Monroe's fossilized turds should not be collected even by those who are big Marilyn Monroe fans doesn't satisfy me as adequate, because it doesn't acknowledge the way we are both drawn to it and disgusted. I'm also not sure the logical argument would be valid. If her face and breasts are sexy, why exactly is her excrement not sexy? Everybody who likes to kiss loves the kissee's saliva, but what about her nasal mucus? What about her tears? What about her breast milk? What about her sweat? What about eye gunk? At some point, it depends on how much you really love Marilyn. Our sexual response includes a crazy embrace of the most physical, and a separation from the physical, and the sane atitude toward that is laughter. Laughter integrates our disgust and our attraction; sometimes a whiff of sweat or sour stomach can be intoxicating because it reminds you the person you love is really there, but it's funny to realize that.

Where else do we have mixed feelings or incommensurate attitudes? Once we start looking at the world with those lenses, the answer is: almost everywhere. Take death for example. We love ourselves and we love our loved ones, but when they're dead, we stop hugging and kissing them and instead bury them in the ground. Do we love them or don't we? Mystically, we might say they're not their bodies, they're not really dead, so it doesn't matter. But then why were we so happy to hug their bodies when they were alive, and why are we sad now that they're dead? The mystic seems to absent herself from our lives if she truly thinks the event of our loved one's death doesn't matter at all, that it's literally as meaningless as the disappearance of these words when I turn off my computer, and exactly as foolish to cry over the one as the other. Even the claim that mysticism helps us love seems not to make sense on this view: If death is an illusion, why honor the hero who sacrifices himself rather than the murderer who kills him? Is logic our buddy here? Consider the memoir *Landscapes of the Metropolis of Death* by the Israeli historian Otto Dov Kulka, who spent his childhood in Auschwitz. He tells the dark joke that the children in the concentration camp believed they would go to heaven when they died, but also thought that the authorities up there would hold a selection for who got sent to the gas chambers of heaven. Is there a logical response that seems remotely adequate?

While mysticism and logic have a one-size-fits-all approach, in comedy we hear the voice of the comedian. Here are some jokes the comedian Robert Schimmel made about the cancer he fought for fifteen years (before dying in a car accident):

> God only gives you as much you can handle. Well, he might have gotten the wrong impression of me. Because I'm pretty overwhelmed.
>
> A chaplain told me: "Death is not the end." I asked him if he could tell that to my uncle—because he was under the impression that he's been dead since 1976. So much so, that he had a funeral, and has been buried since.
>
> A friend told me that death [is] the beginning of a new adventure. Really? Can't tell you how many times I've said, "Hey, I'm feeling adventurous. Think I'll die. Get stuffed in a box

like a pepperoni pizza. Thrown into a hole in the ground. Covered in dirt. And decompose. Anyone wanna come with? (cited in cliffviewpilot.com)

Schimmel's jokes allow us to experience our contradictory attitudes toward death: We think it's the end and we're afraid of it, and we hope it isn't.

His best joke, in my opinion, is this one, which my brother told me when he was diagnosed with leukemia: "My son got cancer and I thought that was really bad. But then *I* got cancer."

The socially acceptable view is that we don't want bad things to happen to other people, and if we get in the habit of saying, whenever our friends get sick or lose their jobs or their loved ones, "Better you than me!" the full force of social disapproval will come crashing down on our heads, and we won't have any friends. Society expects us (especially if we're women) to say we are happy when other people get things and sad when they lose things, and to minimize how happy we are when good things happen to us and not them. We love heroes and saints who meet other people's needs, but if you think about it, that shows how much we like our own needs met. If nobody wanted their needs to be met, then we wouldn't admire the heroes and saints, and in fact, they would be out of a job. It should go without saying, but given that there are people who have cancer and can't afford medicine, every time we take money and buy something (say this book, for example) rather than giving it to those people, we are putting our own needs ahead of theirs.

And yet. We all die sometime, and if all we do in our lives is think about our own needs, it's a pretty limited life. We don't just pretend to care about other people; we actually do. We actually do sacrifice our own enjoyment for our children all the time.

But, on the first hand, how far would we take it? Would we sacrifice our lives to save our children? If our child had cancer, would we take the cancer on ourselves? Some would. And they don't know who they are. If I imagine the scenario, first of all God forbid, and second of all, even though I can imagine myself as a hero, a perfect dad, and a martyr, I don't know whether those fantasies will be played out in real life.

So Schimmel's joke is a really good joke. It puts us face-to-face with our contradiction. We want to say, "You're not supposed to say that! You're supposed to say it's worse for your kid to have cancer than for you to have cancer." But part of us is like "Of course." I'm not a hero who just gave my life to save somebody. The best I ever did was give bone marrow to save my brother, and that was no worse than three days with the flu. If I had just given up my life to save my child—say by agreeing to donate my heart to treat his heart cancer—I might not laugh at Schimmel's joke. But then again I might, because the hero or the martyr is not just a logical robot who picks the better choice rather than the worst without a care in the world, or a mystic who allows himself to be killed because he knows we are all one so it doesn't matter. As Rodin shows us in *The Burghers of Calais,* the hero struggles. That's why he's heroic. He gets in there and deals with the contradiction between his love of himself and his love of his child. So I think, even more than the rest of us, the hero would enjoy and grant the truth of Schimmel's joke.

Now comes a really deep and important point. As we saw in "A Visit from St. Nicholas," the substance that Santa's belly resembles when it shakes is specifically jelly. This hints to us that another wellspring of contradiction in life is jelly, and a good joke about this topic is Sarah Silverman's. "I was licking jelly off of my boyfriend's penis and all of a sudden I'm thinking, 'Oh my God, I'm turning into my mother!'"*

Okay. I was making a small joke of my own—this joke is not really a joke about jelly. It's a joke about sex.

It's a complex joke with multiple layers, and to explain a joke is to ruin it. But I'm going to ruin it a little, since Silverman can take it.

One aspect of it has to do with mothers. Are mothers sexual? On the one hand, duh, right? Mothers are the most sexual category of the three classical categories of women, the others being, of course, virgins and crones. But on the other hand, do we think about our mothers as sexual? We know that, barring gizmos, they had to have been at least once, but we don't want to think about them that way. We would like it if whenever we call up from our imagination an image of our mother and jelly, the only image emerging would be one involving sandwiches.

---

* Actually, this joke was written by Claudia Lonow and performed by Silverman.

Another aspect of the joke is that it starts off sexy. Robert Schimmel had the impact he had because he really had cancer, and you can't help but imagine him dying, and that makes you sad and scared. Sarah Silverman has the impact she has because, if you're the right kind of straight man or lesbian, she really is sexy, or sexy enough that if you imagine her licking jelly off a penis, it makes you feel aroused. But then she asks you to imagine your mother doing it, and you frantically try to operate your brain so as to turn off the arousal. So the net effect is to take your nervous system on a roller coaster ride. Schimmel is using our actual anxiety about facing a man facing down his own death, and Silverman is using our actual conflicts about facing a woman who is deliberately sexually provocative but then reveals herself to be a self-aware human being who worries that she is falling into the dysfunctional patterns of her mother. We have to regard her in two ways, as a sexual object and an anxious ethical philosopher.

When we're feeling sexy, we're not worrying about whether we're getting old or whether we're repeating patterns, and when we're reflecting on where our lives are going, we're usually not in the middle of sex. But of course it's the same person who licks jelly off her boyfriend's penis and worries about whether she is stuck in the patterns of her mother. We are both sexual and moral, self-conscious beings. We pretend we aren't and we get shocked when, for example, Bill Clinton uses one hand to explore Monica Lewinsky with a cigar and the other to reform welfare as we know it. We reach for moral condemnation to anæsthetize ourselves, but that's just human life; the president liked his tobacco sex and he also liked welfare reform. Anybody with a sex life has a moment when they are more interested in body parts than they are in world peace. If they weren't, it would be highly offensive to their partner and probably be classified as a sexual dysfunction: They'd be unable to be in the moment because they're too worried about world peace. All our parents and their parents, and certainly Moses and possibly Jesus, depending on whom you believe, had moments when they were more interested in sticking their tongues in their partner than they were in ending racism, or ending suffering, or making the world a better place. What's the best response? Is the best response logic? No, that's not very sexy. Comedy lets us embrace our sexuality and also our contradictory attitudes toward it.

8

The Giant Cucumber of Señor Kaboom

So Schimmel and Silverman show us that we can deal with our contradictory attitudes toward sex and death through comedy. Yet not all explorations of contradiction are funny, and not all attempts at humor are bravely facing the contradictions of the human condition. The first problem, contradictions that aren't funny, is the easier one to deal with. I could grab a computer keyboard and say, "Hey, look, it's a chicken sandwich—let me have some hot sauce for this chicken sandwich!"—or I could take a chicken sandwich and start typing on it and say, "Check it out! I'm typing an e-mail! Send!" And it wouldn't be funny; it would just be weird. It would be unfunny because nobody cares about that issue—nobody is emotionally invested in the thesis that a keyboard is or is not like a chicken sandwich. If you try to fill out the situation to make it funny—by imagining, say, we're a bunch of really, really hungry computer nerds who are forced to nibble on our keyboards—you will find you have put in some substance that people actually have conflicting feelings about.

What about the bully who grabs your hand and makes you hit yourself and then asks, laughingly, why are you hitting yourself? Let's admit that it's funny to him—he finds there's something that tickles him about victims who claim that they don't want to be hit being forced to be complicit in their own brutalization. We don't think it's funny, because we're not bullies. The push and pull between the rights of victims and the joy of victimizing just isn't a living paradox for us. That might not mean that comedy is bad, but just that people are able to take even the best things and use them to bad ends. We can use charity to humiliate, language to lie, sex to shame, so of course we can use comedy to anæsthetize ourselves to the suffering of others.

Just as often or more, though, it is the moralists who grab the high ground and are shocked by the comic's heartless humor, who have anæsthetized themselves. When Silverman gave a TED talk, Chris Anderson, the organizer, repudiated her and said, rather ungallantly in my opinion, that her talk was "godawful." Later on, Anderson backpedaled an inch and said she was an acquired taste but not one he wanted to acquire because she was "making jokes about retarded people." What was the joke? It went like this:

> I am going to adopt a mentally challenged child. That's threefold: [one], it's harder for them to get adopted, I have,

> [two], oodles of love to give, and three, I just really enjoy the company of the mentally challenged. There is one caveat I realized, in my plan, and that is, they don't leave the nest at eighteen. You die, God willing, of old age at eighty, and you are worried about who's going to take care of your sixty-year-old retarded child. That's where I came up with this brilliant solution: I am going to adopt a retarded child who is terminally ill. I know. Now you're thinking, who looks to adopt a retarded child who's terminally ill? An amazing person.

On the surface, Silverman is making fun of a certain kind of vain celebrity who adopts children in order to boast about it. On a deeper level, though, she is breathing life and awareness into a contradiction at the heart of our attitude toward being good.

After all, if she were really going to adopt a terminally ill child, she *would* be an awesome person. Wouldn't she? Even if she did it for the wrong reason, if I were myself a retarded, terminally ill child, I would take it. She probably has a nice house, food in the refrigerator, maybe a swing. It's true that some people might say that it would be better to have the virtue of humility on top of the virtue of compassion, but the correct answer to this is: So? Even supposing this were the case, it doesn't follow that compassion without humility is worthless. It's better to be good at tennis and have great skin than just to be good at tennis, but even if you have bad skin, it's still good to be good at tennis. If we had to choose between someone adopting a retarded child and being proud and not adopting the child at all, wouldn't we pick proud and adopting over proud and not adopting? And are we one hundred percent sure that humility is a virtue? If adopting a terminally ill, retarded child is an awesome thing to do, why shouldn't the person doing it recognize how awesome it is? We usually like people to be correct about things. Why in this one instance do we want them to be wrong, and about a topic as important as their own moral worth? Adopting a terminally ill, retarded child is going to be hard; maybe the realization that what she is doing is awesome will keep her going. We like our heroes humble so they don't make us feel bad about ourselves, but if they're too humble, they will never have the stick-to-itiveness to be heroes. It's hard work taking care of a mentally retarded child. And although it seems offensive for her to shop for adopted kids as if they're kitchen appliances, why? Should we be *less* careful picking a kid than a blender? The pragmatism that she (or her character) displays when she anticipates what will happen to her potential mentally retarded adoptee when she gets old is necessary to do anything. It's noble for a nun to leave her comfy convent and go to death row and counsel inmates, but she has to know enough to get into her car to do it. If she tries to get in a banana and

drive over to the prison, she's not going to counsel anybody. If you're going to be helpful, you need to know the score.

Is Sarah Silverman (or her character) being callous to the suffering of retarded people? Even assuming that they are suffering, which is debatable, what would be a better attitude? Crying? Would retarded people like being around a crying comic? The actual attitude most people display is just to pretend they don't exist. What do we think about retarded people anyway? Would we be happy if we had a retarded surgeon operating on us? No. Would we go to a retarded marriage counselor if our marriage were in trouble? No. Although the organizer of TED is not in favor of "making fun of the retarded," to date he has not invited a single retarded person to give a TED talk. So how much does he really like them? I bet some of them would love to go! Walking around on a stage, having people clap, saying a bunch of nonsense—who wouldn't like that? Our attitude toward mental retardation and being smart is quite conflicted. If we have a boss who is an idiot, we will freely make fun of that fact with our coworkers, but not if the boss has cancer, even though the cancer might be his fault—if he ate too many burgers or chose to live in a city with bad air quality—but being stupid almost certainly isn't. Yet although we can make fun of our boss for being stupid we can't, or we're not supposed to, make fun of him for being literally developmentally disabled. But surely that's just a question of degree! We've all had bosses who skip merrily down that fine line.

We have contradictory feelings about the retarded, and we

have conflicted attitudes about whether we want acknowledgment for being good people. Jesus accused the Pharisees of giving charity publicly, but by telling his followers to give charity in secret, he essentially leaves it open for them to be proud of their charity and also their humility. Silverman's routine is shocking in the same sense that waking up is shocking. In the face of the contradictions of life—for example, we want people to think we're good but don't want to be the kind of people who want people to think we are good—we care about retarded people, but we don't want to think about them—the usual response is distraction and numbness. Silverman's comedy brings feeling to both sides of the position, without papering over the pain with a neat verbal formulation or tidy ethical compromise. A television executive once said to me regarding some comedian's joke about rape, "What kind of world does he think he's living in where people think jokes about rape are funny?" In point of fact, the answer to her question is: this one. Witness Sarah Silverman's joke: "I was raped by a doctor, which for a Jewish girl is bittersweet." The organizer of TED said, "Jokes about the retarded are an acquired taste and not one I want to acquire." Why do people have the attitude that it's bad to make jokes about death and sex and suffering? Well, obviously, because they have no sense of humor.

If you laugh cruelly at someone else's suffering, you're a cruel person; if you chuckle with sympathetic pain, you're okay. A friend of mine's favorite joke is "There are two kinds of people in the world: people who don't cry when they're happy and people who do. People who don't cry when they're happy are sorry for the people who cry because they're crying. People who do cry are sorry for the people who don't because they've never been happy." So maybe the people who don't laugh at things that are inappropriate have never really laughed.*

---

* By the way, our relationship to IQ is quite self-deceitful—we claim not to care about it and to be democratic but really care about it deeply, as evidenced by the fact that much of our enjoyment of books and television and movies is simply a way of bragging that we are able to understand them. Saying "I just read the latest George Saunders book" is the intellectual equivalent of saying, "Yeah, just ran a marathon in three hours. No biggie." Or "I just got a twenty-thousand-dollar bonus." Not this book, though—you should enjoy this because you love me.

P. T. Barnum was famous for saying nobody ever went broke underestimating the intelligence of the American public—a lot of people have gone broke in Hollywood since then, so I imagine somebody has, but Kaplan's Correlate is that nobody has gone broke convincing moderately intelligent people that they are very intelligent. So that is one purpose of this book—it's basically a pretty boneheaded thesis—life is hard to understand so you should laugh about it—but it's dressed up with a lot of stuff about Russell and mysticism so you can brag to your friends about enjoying it. The irony of it is that, although I'm really smart, I'm also extremely humble! You may have been annoyed and thought I was showing off, but it's not true—I've been reading philosophy since I was a little kid as a way of dealing with my own emotional problems. So it's nothing to brag about that I've read more philosophy than you might have—it's just my own particular issues, like them or not. I used to pretend to be dumb so people would not hate me, but I finally decided it was too much work.

But there's another, more insane reason why this book is not me showing off. My older brother, Andy, had Down syndrome. That's the reason my

A famous episode of *The Mary Tyler Moore Show* by David Lloyd, "Chuckles Bites the Dust," is a meditation on our conflicting attitudes toward humor. Chuckles the Clown dies when he arrives at a parade dressed as the character Peter Peanut, and a rogue elephant tries to shell him. Lou Grant and Murray Slaughter laugh and Mary is shocked: "A man has died." As an uptight Protestant woman, Mary has been raised to have the proper attitude toward somebody else's misfortune. It seems inappropriate and cruel to her to laugh, and she scolds her coworkers for being insensitive. Lou explains to her, "We laugh at death because we know that death will have the last laugh at us." Lou's message sinks in and Mary realizes she was wrong to look down on her friends. Later, at the clown's funeral, a minister pays tribute to the dead Chuckles in the bland sweet manner of those who are both pro-humor and entirely unfunny:

"Chuckles the Clown brought pleasure to millions. The characters he created will be remembered by children and

---

house was so sad, as I mentioned in the beginning of the chapter on mysticism—I didn't know you so well when we were on page 73, and do now, so I can trust you. Now, here's the thing that's weird and might make you think I'm crazy. In some mystical way, I absorbed his soul after he died—I feel like he is a secret companion inside me. So, consequently, I have the open emotionality of a Down syndrome person while I also have the cognitive resources of a philosopher. I think what I am essentially is a very high-IQ retard.

adults alike: Peter Peanut, Mr. Fee-Fi-Fo, Billy Banana, and my particular favorite, Aunt Yoo-Hoo."

At this point, Mary is stifling laughter, remembering that Chuckles would end the show dressed as Aunt Yoo-Hoo by turning to camera and revealing "The End" written on his bloomers. Earlier in the episode, Murray had suggested they bury him that way.

The reverend continues:

> And not just for the laughter they provided—there was always some deeper meaning to whatever Chuckles did. Do you remember Mr. Fee-Fi-Fo's little catchphrase? Remember how, when his archrival, Señor Kaboom, hit him with a giant cucumber and knocked him down, Mr. Fee-Fi-Fo would always pick himself up, dust himself off, and say, "I hurt my foo-foo"? Life's a lot like that. From time to time we all fall down and hurt our foo-foos. If only we could deal with it as simply and bravely and honestly as Mr. Fee-Fi-Fo. And what did Chuckles ask in return? Not much. In his own words, "A little song, a little dance, a little seltzer down your pants."

Mary struggles to keep from laughing during this lame speech. Finally the minister notices and calls her on it:

> You feel like laughing, don't you? Don't try to stop yourself. Go ahead, laugh out loud. Don't you see? Nothing could

have made Chuckles happier. He lived to make people laugh. He found tears offensive, deeply offensive. He hated to see people cry. Go ahead, my dear—laugh.

And she bursts into tears.

"Chuckles Bites the Dust" shows us the problem with writing a philosophy book about comedy. Once we realize that comedy may in fact be an appropriate response, it stops being inappropriate and we can't laugh.

It's kind of funny also that the whole point of a funeral is to let people feel their grief and yet it's so public that Mary is too embarrassed to feel hers. Why do we even bother having them? Why do we have ceremonies of grief that we don't feel comfortable feeling grief at? For that matter, why do we have ceremonies of sexual love that make it unsexy, or write books about comedy that make it less funny? The Monty Python cheese shop sketch thematizes the contradiction between our social roles and what we really want. It starts with an upper-crust customer asking the middle-class proprietor of a cheese shop for cheese. As he mentions different kinds, the cheese shop owner keeps saying he doesn't have them until we start to suspect that maybe, although he is running a cheese store, he doesn't have any cheese at all. The exchange continues:

Customer: Venezuelan beaver cheese?

Owner: Not *today,* sir, no.

The owner may be tipping his hand by not responding to this clearly insane request. It is as if he knows he is just playing a game with the customer. But he doesn't let the customer know he knows, and by not letting him know he knows, he does let him know. So the customer feels he is making some progress.

Customer: (*pause*) Aah, how about Cheddar?

Owner: Well, we don't get much call for it around here, sir.

Customer: Not much ca—it's the single most popular cheese in the world!

Owner: Not 'round here, sir.

Customer: (*slight pause*) And what *is* the most popular cheese 'round hyah?

Owner: 'Illchester, sir.

Customer: *Is* it.

Owner: Oh, yes, it's staggeringly popular in this manor, squire.

Customer: Is it.

Owner: It's our number one bestseller, sir!

Customer: I see. Uuh . . . 'Illchester, eh?

Owner: Right, sir.

Customer: All right. Okay. Have you got any? (*expecting the answer no*)

Owner: I'll have a look, sir. . . . nnnnnnnnnnnnnnnnno.

Customer: It's not much of a cheese shop, is it?

Owner: Finest in the district!

Customer: (*annoyed*) Explain the logic underlying that conclusion, please.

Owner: Well, it's so clean, sir!

Customer: It's certainly uncontaminated by cheese. . . .

Owner: (*brightly*) You haven't asked me about Limburger, sir.

Customer: Would it be worth it?

Owner: Could be. . . .

Customer: Have you—SHUT THAT BLOODY BOUZOUKI OFF!

Earlier in the sketch, the customer was asked if the bouzouki music bothered him and he said it didn't, he in fact "delighted in all manifestations of the terpsichorean muse." The inane music has been continuing during all the previous, contributing to the irritation until he finally blows up.

Owner: Told you, sir. . . .

Customer: (*slowly*) Have you got any Limburger?

Owner: No.

Customer: Figures. Predictable, really, I suppose. It was an act of purest optimism to have posed the question in the first place. Tell me:

Owner: Yessir?

Customer: (*deliberately*) Have you in fact got any cheese here at all?

Owner: Yes, sir.

Customer: Really?

(*pause*)

Owner: No. Not really, sir.

Customer: You haven't.

Owner: No, sir. Not a scrap. I was deliberately wasting your time, sir.

Customer: Well, I'm sorry, but I'm going to have to shoot you.

Owner: Right-o, sir.

(*The customer takes out a gun and shoots the owner.*)

Customer: What a *senseless* waste of human life.

This is obviously about British reserve and class—a man who is too polite to get angry at the middle-class proprietor, and a middle-class proprietor who can't explicitly tell the customer to go to hell but instead uses social conventions to passively aggressively attack him and deliberately waste his time. But of course that's not what it's about. Because cheese is milk—it means infantile satisfaction. And all social arrangements,

British and American, class-based and relatively class-free, frustrate us. We want the milk and we never get it and we go through the charade when deep down we're so frustrated we want to kill everyone because we're never, never happy.

The cheese shop sketch is obviously about the Holocaust and the destruction of European civilization in World War I and II.

Thomas Mann's *The Magic Mountain* compared prewar Europe to a tuberculosis sanatorium full of people arguing about which is better, faith or reason. The cheese shop sketch goes deeper: It says that European civilization was a dance of death between people trapped in roles they didn't believe in anymore. The owner knows he has no cheese; the customer knows he will not get any cheese. It doesn't make anybody happy, but people do it because they do it—priests preach about a God they don't believe in; kings run around executing people even though nobody, including them, believes in kings; everybody goes through the motions until finally people can't take it anymore and kill each other.

Really? No, it's much deeper than that. Because in the version on the record (this version I got off the Internet), the cheese shop owner says no, he doesn't have any cheese. And the customer says, "All right, I'm going to ask you again, and if you say no, I'm going to shoot you." And the owner says "right," and the customer asks him, the owner says no, he has no cheese, and the customer shoots him.

Is Monty Python advocating violence as a solution? In a

slightly different version of the sketch, found on the *Monty Python Matching Tie and Handkerchief* album (1973), the owner admits to deliberately wasting the customer's time even though he knows the end result is the customer will shoot him.

Customer: Tell me something, do you have any cheese at all?

Owner: Yes, sir.

Customer: Now, I'm going to ask you that question once more, and if you say no, I'm going to shoot you through the head. Now, do you have any cheese at all?

Owner: No.

Customer: (*shoots him*) What a senseless waste of human life.

Why does the owner admit he has no cheese? Maybe at the end, he prefers honesty and death to continuing a lie. Or maybe he is so stupid that he just can't imagine he'll actually be killed. Or maybe he opened a cheese shop that didn't sell any cheese because he wanted to die exactly that way, uncontaminated by cheese.

The issue of laughing at things that are inappropriate reveals why comedy is a cousin of mysticism and says what it has to say without falling into its traps, or maybe, by falling into its traps on purpose. Not to be pretentious (or at least to be pretentious in my own authentic way that I can be caught on), comedy is infinite. The second we come up with a justification for laughing, it stops being funny—it becomes embarrassing. And that's funny. Logic hopes to get the whole job of being a person finished once and for all. Someday we will know all the facts and we'll be able to say clearly what we think about them. Logic envisions an end. But comedy can never end. Once you formulate your theory of humor, it will always be possible to laugh at that theory. In fact, it is not just possible, it's a good idea. Once we say, for example, "I get it. I thought it was inappropriate to laugh at Sarah Silverman's joke about retards and Schimmel's joke about his kid's cancer, but now I realize it's appropriate. Sarah Silverman actually likes retards, and Schimmel is only kidding—he's not happy his son has cancer. I've learned from reading this Kaplan Santa book that offensive comedy is not actually offensive—it's just a more sophisticated way to approach paradox." Once you say that, we can then formulate another joke. "I'm not actually making fun of retards, but try explaining that to a retard." Once we've neutered comedy by explaining why it's appropriate, we always still have the option to open our minds a little further and be inappropriate again.

I think it's ridiculous that I've taken funny jokes, explained them, and ruined them. The explanation isn't funny, except that it is funny that I would take a book about Santa Claus and cram it full of philosophy and then jokes about cancer, turds, jelly, and penises. Or it was until I said that.

9
Pain Is Comedy Minus Time

As luck would have it, the television show *The Big Bang Theory* has a scene that deals explicitly with one of the themes of this book: whether and to what extent science and rationality can explain us and our lives. Sheldon Cooper, the self-obsessed, emotionally chilly physicist, and his girlfriend, Amy Farrah Fowler, the neuroscientist who is in love with him, have an argument about the relative explanatory power of physics versus neuroscience. They join their friends in the cafeteria at Caltech:

Sheldon: Greetings.

Leonard: Hey.

Sheldon: I brought Amy here to show her some of the work I'm doing.

Amy: It's very impressive, for theoretical work.

Sheldon: Do I detect a hint of condescension?

Amy: I'm sorry, was I being too subtle? I meant compared to the real-world applications of neurobiology, theoretical physics is, what's the word I'm looking for? Hmm, cute.

Leonard and Howard together: Oooh!

Sheldon: Are you suggesting the work of a neurobiologist like Babinski could ever rise to the significance of a physicist like Clerk Maxwell or Dirac?

Amy: I'm stating it outright. Babinski eats Dirac for breakfast and defecates Clerk Maxwell.

Sheldon: You take that back.

Amy: Absolutely not. My colleagues and I are mapping the neurological substrates that subserve global information processing, which is required for all cognitive reasoning, including scientific inquiry, making my research ipso facto prior in the *ordo cognoscendi*. That means it's better than his research, and by extension, of course, yours.

Leonard: I'm sorry, I'm-I'm still trying to work on the defecating Clerk Maxwell, so . . .

Sheldon: Excuse me, but a grand unified theory, insofar as it explains everything, will ipso facto explain neurobiology.

Amy: Yes, but if I'm successful, I will be able to map and reproduce your thought processes in deriving a grand unified theory and, therefore, subsume your conclusions under my paradigm.

Sheldon: That's the rankest psychologism, and was conclusively revealed as hogwash by Gottlob Frege in the 1890s!

Amy: We appear to have reached an impasse.

Sheldon: I agree. I move our relationship terminate immediately.

Amy: Seconded.

Sheldon: There being no objections . . .

All: No, uh-uh.

Sheldon: The motion carries. Good day, Amy Farrah Fowler.

Amy: Good day, Sheldon Cooper.

Howard: Women, huh? Can't live with them, can't successfully refute their hypotheses.

Sheldon: Amen to that.

This scene grapples with a contradiction. On the one hand, it seems that physics should explain biology. On the other hand, it seems that biology should explain physics. Intellectually, we shuttle back and forth. But the approach of the writer—me—is comedy. It has something in common with mysticism because it does not try to defuse the paradox—it accepts that falling into paradox is part of life. But it has something in common with the logical approach because it doesn't ask us to just revel in paradox—we get an outside perspective that criticizes both of the paradox's horns. We see that both Sheldon's and Amy's perspectives are limited because they are ignoring another level of what's going on—namely that they are a boyfriend and girlfriend having a fight about what the terms of their relationship are going to be, and how they can maintain their individuality while committing to a relationship.

Another thing this scene demonstrates, I hope, is that comedy, when it works, causes pleasure. Maybe the pleasure comes from the release of tension that builds up as our minds bounce from pole to pole of a contradiction, or maybe it is the pleasure of enjoying the bouncing. However you want to say it, comedy lets us live in, accept, and integrate two parts of ourselves.

This episode ends with Sheldon and Amy patching up their differences and forming a relationship. And this is traditional. Since ancient times, comedies have always ended in a wedding. This is because they are the formation of larger wholes—both between people and between warring subsystems within the self.

Comedy and laughter are linked at a neurobiological level to the play system, a part of the brain that the neuroscientist Jaak Panksepp has found exists at least as far down the evolutionary scale as rats. Rats enjoy being tickled, enjoy playing, and even

laugh.* What kinds of contradictions do rats have to laugh at? Probably the contradiction between "I am going to bite that rat" and "That rat is going to bite me." Play is a joyful engagement with others, but it is one in which relations of top dog and

* Rat and dog laughter doesn't sound like human laughter, but it can be identified as laughter because the animals seek situations that bring it about, and they get put in a good mood by listening to it.

bottom dog (or top rat and bottom rat) rapidly switch. In fact, scientists have studied this, and if two kids or two rats are play-fighting, one will come out the victor no more than 70 percent of the time. If you do win more than 70 percent of the time, then it's not fun anymore—it's bullying.

In comedy, we play with each other—switching who will be the boss of whom, and who will win and who will lose. And we play with opposing ideas and opposing horns of a paradox. The sudden switching from top to bottom and bottom to top makes us laugh. If we do it right. If we don't, it's just intellectual bullying: me making you hit yourself with your own mind and then asking why you don't stop.

Unlike logic and like mysticism, comedy embraces contradiction. It unifies the two sides of a paradox into a larger, livable whole without denying either side. Like logic, and unlike mysticism, comedy is antiauthoritarian.* It points out contradictions and gives us tools for criticizing them. But compared to logic, comedy encourages us to take a compassionate and forgiving approach to our limitations. So comedy has the potential to bring about a healing of the warring approaches of mysticism and logic.

If part of the task of becoming human is integrating emotion and cognition, heart and mind, it would make sense that the dialectic we've just gone through on the intellectual side has a parallel on the emotional side. In other words, if we're

---

*** Possible counterexample to the authoritarian nature of mysticism: Taoism's influence on the Yellow Turban Rebellion.**

telling the story of a divided self that seeks integration, then this story is actually two stories. The one we have been telling is about how a mind comes to deal with contradiction and paradox and thereby comes to connect with its emotions and own more parts of itself. The very same story told by the estranged emotions is the story of how we come to deal with trauma. And it also ends with comedy.

The psychologist Mary Ainsworth developed an instrument to diagnose attachment disorders, called the Adult Attachment Interviews. This instrument finds that an ability to recount past traumas in an accurate and gently humorous way is the chief diagnostic marker of psychological security.

The clinician administering the interview asks patients to look back on and describe traumatic events in their childhood. The two main categories insecure people fall into are "dismissing" or "preoccupied." The dismissing deny that anything interesting happened to them. They say things like "It was normal!" "Fine!" "Great parents—what do you expect me to say?" The preoccupied people are suddenly back there with the trauma—"My mother said I looked fat and it's not my fault! I'm just nervous! But you keep calling me fat! And then you feed me! What do you want from me?" To me it seems that the preoccupied have something in common with the mystics—they don't make any rational sense, while the dismissing have something in common with the logical—it's as if their lives are happening to someone else, and they are looking at it all as a problem to be solved, from very far away. Both of them, the overly logical and the overly mystical people, are unable to describe their trauma in a realistic and humorous and self-forgiving way. They not only have attachment disorders but statistically tend to raise children who will have attachment disorders. As the psychiatrist Daniel Siegel points out, the task of the attachment interview is to balance autobiographical memory, which is a right-brain function, with talking, which is a left-brain function. It's a task of hemispheric integration. The preoccupied person is flooded by input from the right brain—in the moment, emotional, contextual—and unable to relate it to the speech situation he is in. The dismissive person answers the question using his left brain exclusively and is totally cut off from his right, so he has no access to emotional, autobiographical memory.

The intellect wants to understand, so its rupture falls between what we understand and what we don't understand, what we believe and what we can't believe. In our emotional lives, on the other hand, the paradox we need to overcome is that between safety and danger. We need to be safe, but we know we aren't, and the fundamental task we are faced with is to achieve a point of view that says that we are safe enough to explore the environment but that takes into account the real dangers. Are we safe or are we in danger? It's not a question that admits of a cognitive, detached answer. Or, rather, every way of living a human life is an attempt to answer this question—from the stoic's withdrawal to the hysteric's quivering vulnerability. The comedic answer allows for joy, growth, and forgiveness, of ourselves and of others.

The same phenomenon that the intellect calls "contradiction" is called "trauma" by the emotions. We feel the conflict between our expectations and the painful reality that overwhelms them.

We can view our history as a species as the autobiography of a single human with a very long life. Not just mystically but concretely, because culture is an attempt to allow learning to transcend the individual lifetime, in a sense converting our sense of history into a collective memory.

If you view human culture as an attempt to learn across lifetimes, then the fact that by and large our approach to life has been split between the two warring approaches of logic and mysticism means that we as a culture or as a species that can remember its past suffer from what psychologists call disorganized attachment. Children grow up with disorganized attachment when they've been abused. The very thing that we attach ourselves to is the thing that hurts us, so we're simultaneously driven to run toward it and run away from it. It's impossible to do both, and crazy-making to try. So it makes sense that if we reflect, we will get torn in two directions—which is what I have been calling paradox, and have suggested healing through comedy.

At the end of his life Sidney Morgenbesser asked, "Why is God punishing me? Is it only because I don't believe in him?" Whatever our source of ultimate security is—whether it's God or other people or the human capacity for good—if we've grown up at all, it's broken our heart many, many times. Comedy is one way to forgive ourselves and put our hearts together again and maybe, if we're lucky, a little bigger.

PART 4

LIFE

10
Reduced to an Absurdity

But—does Santa Claus exist?

I argued that life is fraught with contradiction and that of the approaches we've studied—logic, mysticism, and comedy—comedy is the best. Comedy integrates different parts of us and allows us to forgive differences we have with each other. It causes pleasure and helps us grow from rigidity to creativity. Comedy lets us know how to approach the unavoidable contradictions in our life. Santa Claus is, for some people, one such area of contradiction. So if we want to know whether or not Santa Claus exists, we need to look at the world with a sense of comedy.

But isn't this a reductio ad absurdum of everything we've said so far? How could whether or not somebody exists depend on comedy?

Well, let's make a couple of qualifications.

First, the comedy that lets us know what exists isn't all comedy. There is cruel comedy and frivolous insensitive comedy and racial humor.* The comedy we need to consult is *good* comedy. And second, comedy isn't the only approach to life that lets us know what exists. There might be others—awe, maybe, or tenderness, or maybe "the dance"—but good comedy does the job.

But even with these qualifications—how could it be that what exists depends on what is funny? Isn't what exists sort of—*harder*? I mean not literally, because pillows exist. But still?

I think the best way out of this puzzle is to notice that your judgments of what exists are the same kinds of judgments you make about how to live your life. There aren't two kinds of things we do: judge what exists and decide what we want to do about it. Fundamentally, there is one kind of thing we do—live our lives—and we can reflect on this activity more or less abstractly. So, supposing my wife yells at me for coming home late, and I yell back and say she yells too much. If I abstract back one step, I can ask myself, "Is this the kind of fight I want to have with my wife?" If I abstract one step further, I can say, "Was it fair of me to say she yells too much? Was that a low blow?" And if I abstract a few more steps, I can wonder whether "fairness" or "low blows" exist. But whether I'm right up against the fight, with my wife's hot breath on my neck, or hovering

---

* In my view, the people who are best at avoiding the pitfalls of racial humor are the Finns.

somewhere up way above it, wondering about fairness, I'm still living my life and judging it.

So it is not absurd that judgments of existence connect up with judgments of what's funny. It makes sense. And it makes sense that when I started by figuring out how I should relate to Tammi and her son and my son, and I got into a consideration of Santa Claus, that ultimately led me to the issue of what kind of approach I wanted to take to contradiction, and that in turn to judgments about what I think is funny. They're all examples of me grappling with my contradictory life at more or less abstract levels.

We can get a better grip on this idea by investigating what it would mean to disagree with it. Suppose a philosopher, Q, argues that what exists does not depend on whatever kind of fight I'm having with my wife or my next-door neighbor and the attitude I take toward it, because what exists depends on science; and if you don't believe that, or understand it, you're dumb and/or kidding yourself. On this view, only the entities we refer to in our most up-to-date physical theory exist. So, according to Q, Q himself doesn't exist, meanings don't exist, and good and bad don't exist. The only things that exist are quantum fields. Who is right? Me or Q?

First of all, Q's view is a lot harder to argue against. If you throw in your allegiance with Q and become a Q-ean, Q can teach you a way of arguing so you will never be proven wrong. First of all, ask a scientist what exists in the latest scientific theory. Don't worry, you don't have to be able to understand the theory; you're just going to have to repeat it. Let's assume that he says quantum fields.* Now, if somebody asks you if x exists, if x is a quantum field, say yes; otherwise, say no. So if somebody asks you, "Do you exist?" say, "No. I'm quantum fields." If somebody asks you, "What about your desire to understand what exists?" say, "That's just quantum fields." "What about the conversation we're having now?" "Quantum fields." "Why do you care?" "Quantum fields." In fact, you don't even have to listen to what they say, you can just prepare your response. Which is sort of the problem.

Just because nobody will be able to prove that you're wrong doesn't mean you're not wrong. A proposition's being debate-proof is no guarantee of its being correct. In a famous anecdote related by William James, a psychiatrist is examining a mental patient who is under the delusion that he is dead. The psychiatrist asks him, "Do dead men bleed?" and the patient answers no, whereupon the psychiatrist takes out a scalpel and makes a small incision in the patient's arm. As a drop of blood forms on his skin, the patient looks at it in amazement and finally says, "Well, I'll be damned. Dead men *do* bleed!" People who suffer

---

* This is exactly the procedure I followed in writing this: I e-mailed David Saltzberg, UCLA professor of physics and science advisor to *The Big Bang Theory* and asked him what physicists believed existed, and he answered that they believe in quantum fields. Do I know what a quantum field is? No! I imagine it's some kind of weird fluctuation stretching across the universe and might have something to do with virtual particles popping in and out of existence, but I have no idea.

from autism view others as something like collections of matter, and you can't talk them out of it—it takes therapy. The solipsist claims to believe that he's the only one who exists, but we all know he's wrong. I mean I know I exist even though I can't prove it to him! Even if he annoys me so much with his solipsism that I end up killing him, up to the point that I kill him, he won't believe I exist, and afterward *he* won't exist, so he won't be around to acknowledge I'm right. (Not sure about during.) At no point in our social intercourse will I convince him that anybody else exists. But I do and so do you and so do problems that need solving. I guarantee it!

How can I be so sure? Well, consider if the opposite were true. Supposing you didn't exist. You wouldn't need to do science or figure out what existed or didn't exist, because you wouldn't exist. There would be quantum fields, but quantum fields don't need to know what exist—they just do whatever it is quantum fields do. Probably fluctuate. And if other people didn't exist, you couldn't need to take what they have to say into account, or tell them what you think. And if problems and things to be done and fixed and resolved didn't exist, then you wouldn't feel the need to do science, and you would have no way of getting from less scientific insight to more scientific insight. So at least three things have to exist—you, me, and unfinished situations, such as problems that need solving and work that needs doing.

But does everything exist? That won't work, because sometimes the problems that need solving and the work that needs doing require us to change our view about what exists. Some people used to believe that Shitala the smallpox goddess existed, but those people couldn't protect their children from smallpox as well as people who believe there is no Shitala, and smallpox is caused by the Variola virus. Otto Neurath, the early-twentieth-century philosopher who invented the universal symbolic language that lets us know, for example, that a room is wheelchair accessible also invented an image for our epistemic situation: Neurath's boat. According to Neurath, we humans are like sailors on the open sea, and the boat we are sailing on is made not of planks of wood but of beliefs. As we sail the ocean, we never have the opportunity to put our boat in dry dock, take out all the beliefs that we don't like and replace them with beliefs that are better, but we can rebuild it on the open seas in mid-voyage. So if we become unhappy with our belief in the smallpox goddess because it doesn't help us protect our kids or it makes us give too much money to the priest at the smallpox goddess temple, we can take out the plank in our ship that is the belief that a goddess causes smallpox and replace it with the belief that smallpox is caused by a virus.

So when I said something exists only if it has a context in your life, I didn't mean that you have to rubber-stamp nonsense, or other bad things like racism, because part of the context of your life is self-criticism and change. You don't have to accept that Santa Claus exists if you think he's something that individuals or cultures should grow out of, because healthy people and cultures change. On the other hand, if there's something you think exists and you can't see your way clear to a position of Neurath's boat in which you will ever deny that it exists, then it exists.

So since it doesn't seem that you are ever going to sail to a place where we think there is no such thing as the Earth, the Earth exists. I can't imagine a journey on our Neurath's boat of belief where we are going to throw out that particular plank. So it would seem, according to Neurath, that yes, the entities of science have to exist, but so do people because they're the ones doing the science, so do problems and so do things that are better or worse,* because if you remove any of those planks, the journey stops. If you have someone who takes out the belief "Anything is better than anything else," he doesn't continue doing anything. He dies, literally, because he doesn't believe that eating is better than starving.

It seems as if Neurath can tell us how we will go on a journey that ends with a belief in viruses and not smallpox goddesses. Since you care about not getting killed, we have to believe that dangerous physical situations exist and have to appeal to some notion of causality, time, and space. And therefore, if we want toys, we need to understand that they are made by poorly paid

---

*Is there any difference between believing that some things are better and worse, and believing in the Good? Some philosophers argue that there is a difference, and that it's hugely important. People who believe in the difference call people who believe in better and worse things nominalists, and people who believe in the Good, realists. I've never been able to understand what the difference is or why it matters.

workers in China, not by Santa's elves in the North Pole. So . . . Santa Claus does not exist.

The end.

Except that we are assuming we know where we are going. There have been societies where people have starved themselves to death because they were making a political protest, or believed in a religious or aesthetic goal that is more important to them than physical survival. So what counts as a successful voyage for them will be very different, and what counts as good boat design will be different too. Medieval Christians, at least the serious ones, thought that matter and the body were snares of the devil, so, very logically, they didn't develop a belief in science. Similarly, if you don't care about not being killed, it will be hard to persuade you to believe in the reality of the smallpox virus, and if you care about Christmas cheer and the practice of gift giving more than about maximizing utility, it will be hard to get you to stop believing in Santa.

Neurath's image seems to suggest that we are two things: a pilot and a boat. If we take a close look at it, though, it's hard to tell where the pilot begins and the boat ends. If the pilot has the goals for life, and the boat is the beliefs about what's true, then we tell one kind of story. But if the boat has beliefs about what's true and also about what's important, which seems to make sense, given that goals and beliefs about what's important are subject to change, and have in fact changed from time to time and from place to place, then the metaphor of Neurath's boat seems to lead us in a weird direction. If what exists depends on success and failure, and success and failure depend on our goals, then it seems we get to decide what exists and what doesn't.

We need to take a closer look at the idea that we are two things—a pilot and a boat. The French philosopher René Descartes thought we were two things and developed the doctrine of Cartesian dualism, which is just Latin for "Descartes's idea that we are two things." Descartes got his idea from the observation that two fundamentally different things seem to be true about us. We are free, but we are caused. We are moral agents, but we can be understood by science. Maybe (here's hoping) we are immortal, but we are also, obviously, mortal. It's as if somebody told you about an animal in a box, that it could fly

and it couldn't fly, that it had teeth but it also didn't, and you opened the box and found inside an eagle stapled to a shark. Maybe we humans are contradictory and paradoxical because, like the eagle-shark, we just are two things stapled together! This idea goes back to the ancient gnostics, who believed the body was like a tomb with the soul trapped in it. Descartes's version of dualism, post–scientific revolution, said the two things that make us up are "mind" and "body." Just as the flying eagle is stapled to the swimming shark, so the thinking, free mind is attached to the extended, nonconscious, material body.

But are we an eagle-shark? It's a weird, clumsy idea. Phenomenologically, life just doesn't feel like the yoking together of two different things. I pretty much feel like one thing. I don't say my mind has a headache, I say I have a headache. Descartes thought that it was just a coincidence that feelings of hunger mean a desire for food, but that seems wrong: We can't really make sense of the notion of an experience of hunger that feels just the same as our hunger but means we are full. Even when I do feel split, as for example when I feel torn between anger and my view of myself as a calm person, or I want ice cream even though I'm on a diet, or feel sleepy and also want to stay up and get my book done, none of these feels like a split between a purely physical reality and a purely mental reality. Emotions, moods, and habits all seem to fall between the stools of mind and body. They are both.

Is being a mother a fact about body or mind? When there was no choice in the matter, it seemed like a fact about a woman's body—either she had borne a child or she hadn't. Now that in vitro fertilization, surrogacy, and adoption are possible, being a mother is a question of self-definition, and therefore mental. However, if someone were addicted to adopting children and struggled in vain to stop defining herself through motherhood, we could say that the emotional addiction was in turn functioning as body in relation to the higher level of mind that struggled to overcome the addiction. If we imagine a science fiction technology whereby we are able to alter how many limbs or heads we have, whether we have ovaries or testicles, what electromagnetic spectrum our eyes are receptive to as easily as programming a computer, simply by an act of will and pressing a few buttons, then those features would all be mental.

Phenomenology can mislead, but even if our intuitions were wrong and we were a mind attached to a body, how would the attachment work? Every time you turn a page of this book, a desire, which is mental, is causing you to move your hand, which is physical, and every time a photon bounces off this book and hits your eye, a physical change is causing a thought. We can understand how the shark is attached to the eagle—by stapling—because they're both material objects. How could an immaterial thing—a mind—be attached to a material thing? If there were a glue that could connect the material and the immaterial, then it would have to be both material and immaterial, and if there is such a glue, why not just say we are made of that? Descartes thought the glue was the pineal gland, but everyone agrees that that makes no sense.

Now that we have dispatched the weird and monstrous eagle-shark, let's get back into Neurath's boat.

Neurath pointed out that how well we are getting on in our lives and what we believe in are two variables in a single equation. We solve what we believe by taking into account how successful those beliefs are. But the equation has at least two other kinds of variables in it—where we want to go and who we are. Our lives are not just equations in which we solve for unknowns: We are unknowns as well. The sailor isn't a fixed point in a boat made of changing planks; he or she is made of changing planks too. We are a single reality on the open seas, trying to solve its problems but changing itself along the journey—what it believes, what it wants, and what it is.

And this thing is not a mind-brain hybrid. It is a human brain.

Therefore: The question "Does Santa exist?" is really the question "Do we want to transform our brains from the kind of brains they are now into brains that believe in Santa?"

But how could we possibly answer such a hard question?

11

Santa on the Half-Brain

For a start, we could find people with different kinds of brains and ask them how they like them. How can people have different kinds of brains? Brains can vary along a number of different dimensions. One is hemispheric dominance. Paradox and contradiction and doubleness of belief are not just something we fool around with when we're trying to show how smart we are. There is doubleness built into our anatomy, and paradox built into the architecture of our brains. We have two eyes, two kidneys, two testicles or two ovaries, and two cerebral hemispheres, left and right, connected by a small band of tissue called the corpus callosum. We know, from the testimony of patients who have had a stroke on one side or the other, that these two hemispheres experience reality in profoundly different ways. Jill Bolte Taylor is a brain scientist who suffered a massive stroke that destroyed a chunk of her left hemisphere.

It was an event with pros and cons. On the con side, she could not use language or a phone. On the plus side, she had an experience characterized by overwhelming peace, beauty, and goodness, much like William James's experience on nitrous. Everything sort of flowed together. Once she got her left hemisphere healthy again, she was able to describe her experience, in a TED talk no less:

> Our right hemisphere is all about this present moment. It's all about right here, right now. Our right hemisphere, it thinks in pictures and it learns kinesthetically through the movement of our bodies. Information, in the form of energy, streams in simultaneously through all of our sensory systems. And then it explodes into this enormous collage of what this present moment looks like, what this present moment smells like and tastes like, what it feels like and what it sounds like. I am an energy being connected to the energy all around me through the consciousness of my right hemisphere. We are energy beings connected to one another through the consciousness of our right hemispheres as one human family. And right here, right now, we are brothers and sisters on this planet, here to make the world a better place. And in this moment *we are perfect. We are whole. And we are beautiful.*

This sounds so exactly like the Upanishads that it gives us new insight into our discussion of mysticism. Mystics, it seems, are experiencing a tuning down of the left hemisphere and tuning up of the right. If that's so, it would make sense that logic is a tuning up of the left hemisphere and a tuning down of the right, and comedy is an integration of the two hemispheres. So when we called mysticism to task for fatuousness and authoritarianism, and poor-mouthed logic for not being able to deal with love and spontaneity, what we were really doing was evaluating different brain configurations. Consequently, if you want to know if Santa Claus exists, all we have to do is: First, figure out what the optimal relationship is between the hemispheres, and then, once you've achieved that optimal relationship, see if people with that configuration believe in Santa.

What would the optimal relationship be? In his book *The Master and His Emissary,* the psychiatrist Iain McGilchrist compares the right hemisphere to a king and the left hemisphere to a bureaucrat who works for him. The right hemisphere dispatches the left hemisphere to perform tasks that the left hemisphere is good at: It divides reality into little chunks that are easily manipulable, assigns words and concepts to them, and manipulates them with rules, ignoring the larger context of implicit background understandings. In McGilchrist's view, although this is the optimal relationship, in Western history it has gotten out of whack; the left hemispheric emissary has started to mistake himself for the master. So Scrooge, who doesn't believe in charity because he can't give a utilitarian justification for it, is being ruled by his left hemisphere, while Dickens and the ghosts he imagines are calling people like him back to the importance of the right—cooperation, forgiveness, generosity, and the specialness of the season.

What do the two hemispheres have to say about Santa Claus? The left hemisphere believes everything either exists or does not exist and, since it has never run into Santa, concludes he does not exist. Does the right hemisphere think Santa does not exist? Well, it certainly doesn't say, "Santa Claus exists" or "Santa Claus does not exist," because talking in an abstract way is a left-hemisphere function. But it can sing along to Christmas songs about Santa and participate in Santa-based rituals. The right hemisphere really doesn't take a stand on whether or not things exist any more than when we are in a dream we say

whether or not the things we are involved with exist. We just dream them. Another example of right-brain ontology is our stance toward what D. W. Winnicott calls transitional objects—for example, the child's teddy bear. The child does not assert either that the teddy bear is real or that the teddy bear is just a toy. The child is not a metaphysician. However, the child does talk with the teddy bear, talk as the teddy bear, miss the teddy bear, and play with the teddy bear. He deals with the teddy bear, and its existence doesn't come up.

Instead of saying the gods exist, the right brain might instead use an expression my Thai monk friend Chalor Kosadhammo once used about them: "They are available."

Not only are the answers the left and right brain give different, but the way each approaches questions is different as well. If we are dwelling in our left brain, we are going to try to make an inventory of what exists by making abstract statements that are true at all times. So, for example, we will ask, "Does Santa Claus exist?" and, to answer it, we will try to take the meaning of the statement "Santa Claus exists" and see if it applies across different moments. Then, once we have determined which abstract statements apply to all moments, we will be left with the problem of how we feel about them, and what to do.

It's obvious why this procedure will not work for the right brain. Since, for the right brain, only *this* moment is real, the right brain cannot commit itself to believing one statement will be true across all future times. Since the left brain thinks language is a fungible code, we think when in left-brained mode that it makes sense to take a sentence, e.g., "Raising kids is hard," and ask simply, "Is it true or not?" It's right there on the printed page, so surely we can wonder whether it's true or not and challenge ourselves to define the concepts of "kid" and "hardness" and so on. It seems like a meaningful question to ask, but what if it isn't? Maybe it seems meaningful only because the institution of writing allows us to take a piece of language, write it down, keep it, and carry it around from context to context? Suppose I come home after a hard day, sigh ruefully, look my wife in the eye, and with a particular tone of voice and a particular body language—slumping in my chair—say, "Raising kids is hard." I am expressing an attitude toward kids in a particular way and, by doing so, I have conveyed vastly more information about my attitude toward kids than I would have if I had just written the string of marks "Raising kids is hard." The first expression, "Raising kids is hard"—said in a particular tone of voice, with a particular facial expression and body language, and in a particular interpersonal context—is not incompatible with another expression of language, "Raising kids is fun!"—said with a different expression, body language, and interpersonal setting. The two expressions "Raising kids is hard" and "Raising kids is fun" are contradictory and paradoxical only if we turn them into written expressions. In their spoken context, they can get along just fine. The left brain ultimately seeks to engage the world in the language of science and logic—expressions that are entirely free from personal context—a view from nowhere,* and it seeks to formulate a set of true statements that correspond to reality for everyone at all times in all places. For the right brain, this is an inherently quixotic project. Written expressions such as "The world is made out of atoms" or "Democracy is the best system of government" have so much information removed from them—tone of voice, body language, emotional force, and interpersonal coloring—that they are hard to evaluate. While the journey from particular language to abstract language is viewed by the left brain as a journey from vague to precise, for the right brain, it is a stripping away of the very elements necessary for truth and meaning.

The relationship between theoretical and practical reason is different for the two hemispheres as well. Once the left brain has come up with a set of true statements about reality, it is then stuck with the problem of what to do about it—how to "derive ought from is." For the right brain, all linguistic expression, all thought, is already part of a practical, human-lived moment. Just as a cat uses its tongue to lick, and this licking makes it feel good by making it flourish or makes it feel bad by inhibiting its

* In Thomas Nagel's phrase.

flourishing, so it is when a human being uses his tongue to talk. In a particular moment, in a particular context, a particular expression of thought or language is helping us or hurting us. Theoretical reasoning and practical reasoning are both abstractions from the activity or life of the moment.

Then how do you figure out how to get our brains in the best configuration so you can finally figure out if there's a Santa Claus? Imagine yourself at a desk seated across from your two hemispheres, who have come in for a job interview.

Your assignment is to figure out whether to hire either of them to work for you and if so, doing what. The left hemisphere offers to make logical determinations of reality, and the right hemisphere offers to let you emotionally connect with people, feel things, and be in the moment.

The left hemisphere makes some good arguments:

Left Hemisphere: Look, if you don't think carefully, you might die. You don't want to die. Therefore, you should listen to me.

While the right hemisphere talks poetically:

Right Hemisphere: There is a budding morn in midnight.

So how do you decide which to hire? If you interview them by having a logical discussion, you are setting things up so the left hemisphere will win, because the right hemisphere "don't talk so good."

What if you just stare into each other's eyes and see whom you trust?

Then the right hemisphere will definitely win, because you form your emotional attachments using the right hemisphere! (In fact, the success of therapy depends not on the theories of the therapist, or any insight he or she provides, but on the positive linking up of therapist's and patient's right hemispheres.) If you interview the two of them, you will probably find yourself emotionally attached to the right hemisphere but intellectually drawn to the cogency of the left hemisphere's arguments.

So what would make sense? Well, by analogy, supposing you're a divorced mom with kids and you're getting married to a divorced dad with kids, and you want the whole blended family to get along. You would start by doing things together that help you blend: you have a conversation, you sit by the fire, you have a drum circle, you argue, you have a good meal, you play catch, you go hiking together. You might also go through some tough times together—not just intellectually tough times but times that actually are emotionally challenging and a little scary, so when you got through them, you would trust each other. If you want your left and right brain to get along, you would do the same thing. Go through experiences that challenge both, that both enjoy, that require both: connecting with another human being both emotionally and intellectually, for example, or singing and playing the piano, or being funny while still feeling pain. Then at the end of the day, you could get these two guys to work for you and to like each other and trust each other, and you could pop the question: "Does Santa Claus exist?"

Supposing your team building worked, and you're all a team. You could imagine a couple of possible responses:

The left hemisphere might say, "Look, I'm not so good with this kind of stuff. I have so many emotional limitations. I don't personally understand how Santa Claus could exist, but I trust right hemisphere. He totally had my back at that party. He's a good guy." Or left could say, "He doesn't exist," but you could overrule him and say, "That's not the kind of issue you're good at. It's too emotional." Or you could say, "It's not a good idea to ask that question—it just gets us fighting." Or you could say, "Left, why don't you leave the room—we are planning to have some Christmas time with the kids and you're not so helpful." And then you and your right hemisphere could believe in Santa Claus to your heart's content.

So far so good. Except, who is the you in this parable? You don't have a third hemisphere that sits in judgment over the other two. Where is the self that could get these two hemispheres on the same page? This question actually is going to face us whatever the brain science is. So if the science Taylor relies on is completely wrong, as it may be, and in twenty-five years science reports that the brain is actually made not of two hemispheres but of nine grapes, two coconuts, and a banana, we can ask, "Where is the self that gets that fruit bowl in order?"

How can you stand outside of yourself in order to tell what the best brain configuration is?

Maybe we could solve the problem with science. Let's imagine the following thought experiment. After the year of the brain, the decade of the brain, the century of the brain, and, finally, the millennium of the brain, scientists make the announcement: We have finally figured out the brain. Perfectly. That is, they have a theory so powerful it tells us what all the different parts of the brain do and gives us unchecked power to manipulate them. They can hook the brain up to a user interface, so if you press one button on a brain, you make it smarter, another one you make it focus, another one you make it fall in love. You turn a dial and it becomes really, really interested in sports, another and it doesn't care why the New York Knicks just can't bring it this season. Now, suppose they put an experimental subject in there. They put the dial on his head and turn smart up to the maximum and he comes out of the Experimentorium and he says, "Santa Claus exists. And by the way, you should all be my slaves. Kneel before Zoltar!" Assuming his name is Zoltar. If his name is Eddie, he would say, "Kneel before Eddie!"

Would that prove, finally, that Santa exists? Not really, because maybe that smart button was really a smart plus crazy button.* Or maybe the best thing for people, as my mother is

* If this does happen and the super-intelligent supermen read this, I mean no disrespect.

fond of pointing out, is not to be too smart. I imagine if this all transpired in real life, some people would go in and talk to Eddie and then come out and say, "Yup! He is the next stage of human evolution. We should worship him." But if that happened, you would be right to wonder, "Are they being honest or did he just promise to pay them off, say by being his lieutenants in the new World Eddie-ocracy?"

It might end in war: superbrain Eddie and his lieutenants against the rest of the human race! If there were a war and we started to lose it (I'm on your side), would that mean that the things Eddie believes in would thereby be about to exist? Let's say that Eddie believes in Santa Claus and he is about to win the war. Does that mean, then, Santa is about to exist? But what if at the last moment we make a bold counterattack and win! Does that mean that Santa Claus almost existed but then at the last moment, he didn't? That seems very, very strange. If Eddie is a crazy monster and he defeats everybody else on Earth and on top of that he makes us believe weird things, that just seems like bad on top of bad. First, we have to live under a crazy monster, and second, he is going to make us believe weird things. It doesn't mean Santa Claus exists.

Maybe the mistake is making only one Eddie. Maybe if we want to know what exists, what we need to do is make a lot of different kinds of brains. Maybe we should make every possible kind of brain: brains that think in pictures, brains that think in statistics, radical egalitarian brains that accept information from others, and glacially snobbish brains who form their views of life based on their own meticulous mental processes, and maybe we should also include intelligent apes and bears and computers and oak trees, and then we should let them all hash it out. If we had a sufficiently long time, we could create every possible brain and survey them all on what they believed about Santa Claus and, while we're at it, God and the point of life. We could also form all possible teams of all the possible brains and get the teams to submit reports. It would be expensive, but I would be willing to hang out in front of the supermarket and gather signatures for the project, because it would finally answer all our questions.

Or would it? Wouldn't there be too much data? Wouldn't we be back in the Santa Claus/Manta Claus/Mylanta Claus conundrum? It's *possible* to get into those brain configurations, but is it a good idea?

What if we gave them a job? We're all facing down the ultimate end of the universe—maybe we should ask this collection of superbrains to solve that problem and believe the ones who can? But even then (or especially then), we may be faced with irreconcilable differences. Some of the superbrains may tell us there's nothing to be done because the whole universe will start up again, others that it is all a simulation in the mind of a being in another universe, others that it is the end but the journey was worthwhile, others that it's the end but it was all a cruel joke. Just because they're smart doesn't mean they're not kidding themselves, nor does a certain rough-hewn simplicity mean that they are.

Could we let them fight and believe whoever emerged as the winner? What would count as winning or losing a battle royal of disembodied superbrains? Would it be a debate? Or would we just give them monster bodies and have them fight to the death on some alien world? But if we did that, wouldn't this prejudice the situation against those brains who thought duking it out with monster bodies was immoral or just in bad taste?

And yet isn't that the situation we are in? Aren't the beliefs that are currently stalking the brains and exobrains of planet

Earth the victors of previous wars—wars of religion, sure, and wars about whether we will organize ourselves as peaceful hunter-gatherers or Earth-despoiling consumer capitalists, wars about whether or not we will be polygamous or monogamous, whether we will use water clocks or electric clocks or no clocks at all?

12
Does Odin Exist?

The science fiction scenario I got into in the last chapter is a lot like the actual history of the human race. We have been developing different kinds of brains and using our bodies to fight over which brain is best. In one corner stand the Enlightenment and its heirs, who view the perfect brain as that of the autonomous adult who has thrown off the childlike superstitions of the past. In the other corner is its archrival, romanticism, which thinks the perfect brain has to have room for Santa. The romantic thinks that the perfect brain state is a union of innocence and experience. Yes, there's something good about science and autonomy, but there's also something good about the way children look at the world that we shouldn't throw off with disdain. It's great to be a baby suckling at the breast, or a five-year-old playing with pinecones—why lose any of it? Just add the ability to take care of ourselves and others. Adolescents are on a mission to prove that they can take care of themselves, so they brag that they're not children anymore. But the only people who feel the need to brag that they're not children are those who are in danger of being mistaken for them, because that's what they recently were. So maybe the Enlightenment's condescending attitude toward myth is something like the adolescent's dismissal of childhood. A big child doesn't want to be accused of being a diaper baby, but the use of the expression "diaper baby" is pretty immature.

G. K. Chesterton, intellectual godfather to J. R. R. Tolkien, C. S. Lewis, and the current pope, was happy to say both that he believed in Christianity and that it was a fairy tale. He just thought it was a fairy tale that happened to be true. Fairy tales, of course, are pagan myths taught to little children by their babysitters in medieval Europe, who, unlike my babysitter in 1970s Brooklyn, were not hippies but pagans. In the bowdlerized myths of the pagan babysitters, the virgin goddesses become princesses, the father gods become kings, and the monster that eats the sun and causes winter becomes the toad that captures the little girl's golden ball. In pre-Christian Europe, Vikings believed that Odin was the gift giver and portrayed him as a powerful old man who came in the dead of winter. Some see in Santa Claus a contemporary manifestation of Odin.* But is that good news for Santa or bad news for Odin?

* Tor Webster, a Celtic Christian I met in Glastonbury, sees Santa as a contemporary manifestation of Hern the hunter, the strongest warrior of the tribe,

What should the attitude of a mature culture be toward its myths? You might as well ask, "What should the attitude of a mature human being be toward his or her childhood fantasies?" Some might say, painful as it is, we need to give up myths—Santa Claus, Odin, or God. But it is painful, and in the cultural back-and-forth thrown up by the Enlightenment and the counter-Enlightenment of romanticism, you come across a lot of crying.

Here is the romantic poet Wordsworth writing in 1806:

The world is too much with us; late and soon,
Getting and spending, we lay waste our powers:
Little we see in Nature that is ours;
We have given our hearts away, a sordid boon!
The Sea that bares her bosom to the moon;
The winds that will be howling at all hours,
And are up-gathered now like sleeping flowers;
For this, for everything, we are out of tune;
It moves us not.—Great God! I'd rather be
A Pagan suckled in a creed outworn;
So might I, standing on this pleasant lea,
Have glimpses that would make me less forlorn;

---

who would catch a deer in the dead of winter and return wrapped in its bloody skin and covered in snow. According to Tor, that's why Santa is red and white.

Have sight of Proteus rising from the sea;
Or hear old Triton blow his wreathed horn. (Wordsworth 536)

Wordsworth is complaining that modern modes of thought and being make him feel bereft and prevent him from seeing gods. To understand his point, we need to challenge the view that perception works something like a blank screen. On this picture, if you open your eyes and look and your eyes are not impaired, if it's there, it will be projected on the blank screen, and you'll see it, and if it's not, it won't. If I woke up and snuck downstairs on Christmas Eve, I wouldn't see Santa, I would see Dad. So Dad exists and Santa doesn't. One obvious problem with this is that there is plenty there that I don't see—germs obviously, but maybe injustice, and there being three of something—but the metaphor has deeper problems too. If we are going to see Dad, assuming Dad is there, we have to be able to position our eye at the right distance. If the eye is a millimeter away from Dad, we won't see Dad. If it is a mile from Dad, we won't see Dad. If we don't know where Dad begins and ends and don't pick out the surface of his skin, we won't see Dad. And we have to pick out the surface of his skin at a correct level of fuzziness. If it's too blurry, we won't see Dad. We'll see a blur. But similarly, if it's too precise—if we see every fold of skin and every hair and every crevasse in every hair—we won't see Dad, we'll see a horrific hairy wrinkle-monster. And if we are a fish, we won't see Dad—we'll just see a blurry shape (perhaps) through the water—or maybe we don't even see it because we lack a concept of dads. What could a fish see? A human? A predator? A form? Maybe fish see us the way we see clouds, or maybe like the frog, they see more or less nothing unless it's moving, and then they see a blip, as if they were playing a video game and it's 1971. Furthermore, if we think about it, the idea that seeing is like having an image thrown on a screen explains exactly nothing because an image on a screen is an image only if somebody sees it. It's a version of the homunculus fallacy—the fallacy that explains human action by appealing to a little human inside of us who does that action. Compare it to the following explanation of eating: To eat something is to swallow it and bring it to our stomach, where a little man eats it. You see the problem. We also don't even need eyes to see. When blind patients have been outfitted with a prosthesis that connects a camera to a pattern of pins on their tongues, they report seeing their loved ones with it.* And not metaphorically either—they are literally seeing. The eye is less like a screen than it is like a hand. It is able to get a purchase on the look of things, just as our hand is able to get a grip on the feel of things. Vision is a bodily interaction with an environment, not a meter-like registration of physical stimuli.

So it makes perfect sense for Wordsworth to wish he could look at the flecks of foam in the ocean and see Proteus. If he had been raised with the right cultural expectations, he would catch sight of Proteus and hear Triton blow on his wreathed horn. So why doesn't he? Max Weber was more sociologist than poet (99 percent/1 percent), but he also thought the trade of Triton for modern science was a questionable bargain:

> Since asceticism undertook to remodel the world and to work out its ideals in the world, material goods have gained an increasing and finally an inexorable power over the lives of men as at no previous period in history. Today

---

* See Norman Doidge, *The Brain That Changes Itself*, for a discussion.

> the spirit of religious asceticism—whether finally, who knows?—has escaped from the cage. But victorious capitalism, since it rests on mechanical foundations, needs its support no longer. The rosy blush of its laughing heir, the Enlightenment, seems also to be irretrievably fading, and the idea of duty in one's calling prowls about in our lives like the ghost of dead religious beliefs.... No one knows who will live in this cage in the future, or whether at the end of this tremendous development, entirely new prophets will arise, or there will be a great rebirth of old ideas and ideals, or, if neither, mechanized petrification, embellished with a sort of convulsive self-importance. For of the last stage of this cultural development, it might well be truly said: "Specialists without spirit, sensualists without heart; this nullity imagines that it has attained a level of civilization never before achieved. (Weber 124)

Weber thinks modern culture has gone through three stages—polytheism, monotheism, and atheism. In polytheism we experienced Proteus rising from the sea and Triton blowing on his wreathed horn. We lived in an atmosphere that was alive with myth—there was a nymph in the local pond, and when it thundered, the gods were bowling in the sky. Monotheism accomplished the "disenchantment" of the world—kicking out all the nymphs and fairies and replacing them with a single God. Then finally an extreme, Puritan form of monotheism called science turned its weapons on the last myth—God—destroyed it, and left behind a world with no myths. What started out as a religious ideal has ultimately gotten us into a very disgusting, materialistic civilization, and now we can't get out of it. He thinks we have followed Wittgenstein's advice; we've climbed a ladder to a bad place, thrown it away, and are now stuck there.

You might say that we do believe in myths. I once got to hear the old philosopher Quine on a panel growl that psychology is biology, biology is chemistry, and chemistry is physics. Oddly, he had once argued in print that everything we believe in other than sense impressions is a myth:

"Viewed from within the phenomenalistic conceptual scheme, the ontologies of physical objects and mathematical objects are myths. The quality of myth, however, is relative; relative, in this case, to the epistemological point of view. This point of view is one among various, corresponding to one among our various interests and purposes." (Quine 19) Quine is sounding a lot more tolerant here than he really is, because his idea of different interests and purposes was very narrow, more or less limited to a choice between doing math or doing physics. But broadening his point beyond what he actually believed, you can say we do believe in the myth of progress, we believe in the myth that man is able to understand and control nature, we believe in the stock market and a steady job. Myths, like the story of Commonus, have social functions—they're not just optional beliefs. Myths are integrated with rituals, which in turn make us experience time and our bodies in a distinctive way, marking the rhythms of the season with holidays, and imposing communal, soothing rhythms on the body through dance and rocking. Myth-centered societies engage the senses often with a communal meal that knits our bodies together through the shared practices of talking, singing, and eating. Rituals take us out of time to what the Romanian comparative religionist and fantasy writer Mircea Eliade calls *illo tempore*. Every New Year's Eve is chaotic because it reenacts the primal chaos before creation; every New Year's Day is a participation in creation anew; every christening is the birth of the first child. Myths engage the imagination. And when we participate in myth, we have a fluid sense of boundaries between us and the world that Jill Bolte Taylor experienced during her brain hemorrhage—we feel at one with nature. The Santa Claus myth includes rituals: a seasonal celebration of the solstice, with singing, eating, and the receiving of gifts.

Wordsworth is forlorn and Weber bitter about the loss of enchantment. So why don't they do something about it? Why don't they teach their kids to believe in Santa Claus? Weber's word for the historical process that takes us from polytheism to secularism is "*Ent-Zauberung*," which means removing the magic, or *Zauber*, from the world (a "*Zauberer*," for instance, is a wizard). "*Zauber*" is related to the Old English word "*teafor*," which means "red dye," because old magical runes were colored with the magical color red. Why not bring the red back into the world and re-dye it, making it like Santa's costume? Why not believe in Santa Claus?

What do you say?

PART 5

ME AND YOU

13

A Tree House in the Tree of Life

If you want to stretch a muscle, you have to do it gently or the muscle will guard against the stretch and you'll end up tighter than when you started. It's the same thing with your mind. We've been talking about whether or not Santa Claus exists, because that's a gentle way to approach our position as modern people regarding belief. Is it possible to recover a sense of myth? Can we, as modern people, engage with reality in a way that includes ritual and imagination and gives us a feeling of being at home? Can we think that our lives at the end of the day have any point at all?

Everything we've looked into so far suggests that the wrong way to look at this question would be to construct a general theory about reality and the human mind. If reality were the sort of thing logicians think it is, then you could give a general story about what it is to deal with general categories. But if reality is the sort of thing I think it is, then I can't give you a general story. What I can do is tell you how these issues shook out in the context of my own life. You're reading this book, so you're going to have to make the decision in the context of *your* life. I can help out a little because now that you've read my book this far, I'm, to some small extent, part of your life. Hi!

As I mentioned earlier, I grew up in a household of unacknowledged sadness because of my brother's death from Down syndrome the year before I was born. So the idea that nothing had a point hovered in the air even though it was never explicitly expressed, and that made it, obviously, harder to fight against. We didn't believe in God or Santa. I was afraid of my ultimate annihilation, and usually lonely, although I sometimes had a sense of peace or comfort that had nothing to do with anything happening in my life, or any people. I am the opposite of a claustrophobic—claustrophilic, I guess—I like to be in a deep, dark enclosed space, like in a closet behind the winter coats, and then just to be happy that I exist.

There's something intrinsically unfair in me as a grown-up telling the story of me as a child and adolescent: I get to tell you what I think about former me, but former me doesn't get to tell you about present me. I think I'm what my adolescent self naturally grew into—that I understand him better and have achieved those of his goals that were realistic—but he probably thinks I've gone soft in the California sun. In a certain mood, adults regard adolescents as monsters, but adolescents can think the same thing about adults, just as werewolves believe humans are monsters cursed every full moon to remain the same. (They are right, by the way: To were is human.)

Nevertheless, history is written by the victors, and if adolescent me didn't want to grow up, he has nobody to blame but himself. So here is present me's take on adolescent me. Back then, I think he thought he had two problems—finding love and finding a coherent theory of the universe and man's place in it, and he just happened to have trouble achieving either goal. I could find people to love me (I mean I did fine), but I was so unnerved by death and the meaningless universe that it really messed up the relationship. A good relationship requires you to respect the other person who loves you, and how could I respect someone who fell in love with a complicated collection of self-replicating molecules? And even though it, thankfully, was not up to me, why did it matter if the human race or even the universe existed at all? Part of me felt that these things—my life, finding love, the universe—were important, and part of me thought the idea that anything that could ever happen to me could be important was dumb. Looking at the same issue from the cognitive point of view, none of my favorite theories of the universe—materialism, Hindu monism, Buddhism—made my actual life and the people in it (including me) matter. If I asked my girlfriend to run away with me and write novels in New Orleans, how could her response matter if we were both the Primordial Consciousness or the Tao or a bunch of atoms?

From my current vantage point, I think what was going on in my adolescent struggles was a longing to connect. Intellectually, I was trying to connect by forming a coherent philosophical theory of the universe and my place in it; emotionally, I had a longing to connect with people, but the fact that these two attempts to connect were one and the same was hidden from me. I wanted to understand the world and I wanted to be less lonely, but I didn't connect the two wants, because I was disconnected.

I was disconnected horizontally, from my contemporaries, but also vertically, from my ancestors. My mother and my father's mother had both broken away from traditional Judaism as their families made the transition from ghetto life in Eastern Europe to modern America. It was very hard for me to talk to my grandpa Wolf and grandma Gussie, who had been born in Galicia and spoke Yiddish, and if, by time machine, we had had a family reunion going back to the brother of Moses, they would have fit in, but my father and I wouldn't have. Although we lit candles on Chanukah, we didn't believe that a loving intelligent God had created the universe. This was because of what the family had gone through with my brother Andy, but also because every day in the *New York Post*, you could read a story like PET SNAKE ESCAPES AND SWALLOWS BABY, which brought the loving God theory into question.

After college, while I was working as a legal temp in Midtown Manhattan, I would go to the library on my lunch hour, and I came across a chapter in *Major Trends in Jewish Mysticism* by Gershom Scholem. Here was Scholem on Isaac Luria's explanation of how there could be a good God but also a world where my brother Andy had Down syndrome and escaped snakes swallow babies:

> If God is "all in all," how can there be things which are not God? How can God create the world out of nothing if there is no nothing? . . . According to Luria, God was compelled to make room for the world by, as it were, abandoning a region within Himself, a kind of mystical primordial space from which He withdrew in order to return to in the act of creation and revelation. (Scholem 261)

So, according to Luria, my life so far had been lived inside a dark closet inside God: dark because He had turned out the lights so someday the human race could turn them back on. That was an interesting idea. But who was Isaac Luria and why had I not heard of him before?

Who he was was, in a sense, easy: a kabbalist* who lived in

* "Kabbalah" literally means that which is received, and refers to the esoteric dimension of Judaism.

Israel during the generation after the destruction of the Jewish community in Spain, traded pepper, meditated in a hut on the Nile for twelve years, returned to Israel, taught a small group of students for a year and a half in the city of Safed in the Galilee, and then died of cholera along with his family at the age of thirty-eight. I hadn't heard of him for two reasons. For Reform Jews, he was a medieval lunatic, raving about reincarnation, multiple universes, multiple souls, and a weirdly biological view of God—a spokesman for everything they hated about Judaism. For Orthodox Jews, he was the deepest of the deep, a new soul, incredibly holy—they call him the Ari, or holy lion—but also incredibly dangerous. They have a point. His thought inspired such weirdos as the dark wizard Aleister Crowley and through him the aerospace pioneer and magician Jack Parsons and through him the cult leader L. Ron Hubbard. More to the point for Jews, a radical take on the Ari's thought led Nathan of Gaza to embrace salvation by sinning, declare the outlandish Sabbatai Zvi to be the messiah, and lead a third of Judaism into following him, until ultimately Sabbatai converted to Islam to save his life. After this fiasco, the defenders of Orthodoxy put the Ari's teachings under lock and key—you needed to be a married man, with two children, a legal scholar, and at least forty to be allowed to study his thought.

Time passed in the vacated space. I became a comedy writer, got married, and had a family (including a son I named Ari after the Ari), but like the bones in a trick joint, the parts of me that hadn't fit together when I was an adolescent remained unintegrated and would pop apart every now and again. Then, during a time when the empty space was feeling a little emptier than usual, I came across a teacher from Jerusalem who saw himself standing in the direct tradition of the Ari: a transplanted Jewish Syrian Los Angeleno named Avraham Sutton. I read and listened to everything from Rabbi Sutton on the Ari I could, went to Safed in Israel to cry on his grave, stopped driving on Saturday, and started wearing a kippah so large it frightened my family and coworkers. I'm writing this five years later. I'm not in my big kippah anymore and although I don't work on Saturday, I'm much more pluralist, less scary, and more assimilated to norms of liberal American society. You still have the same interpretive problem you had when I was talking about my adolescence: Why do you believe present me and not former me? Maybe Orthodox me was right and now I'm backsliding. But whichever me you want to listen to (sadly, they're all pretty self-absorbed), the Ari is a part of his story and, now that you're reading this, a part of your story too. That's because both stories—yours and mine—are about finding a place for Santa in modern life or, more generally, integrating ancient and modern modes of thought. The Ari is a help with this because all his ideas are hybrids, with one foot in myth and one foot in reason. That means if you try to reduce his thought to something like analytic philosophy, it will slip through your fingers and seem meaningless, but if you jump in with your mind and your imagination and your emotions, it can serve as a tool for getting all these different parts of you talking to one another.

Let's start with the Ari's word for what exists. Luria calls it the Limitless—"*Ein Sof*" in Aramaic, "infinite" in a more Latinate English. We can't compare it to anything else because there isn't anything else—if there were something else, the boundary between it and the something else would be a limit, and it's limitless. We can't conceive of it because we'd have to distinguish it from something which it isn't, and there isn't anything which it isn't. By definition, the Limitless has everything. Is it a collection of matter in empty space? No, because it includes thoughts and imagination, and possibility, and the matrix from which that matter and empty space arose, and number, etc. etc. etc.

Isaac Newton was a student of kabbalah. The English philosopher John Locke reports that Newton explained to him that God created the universe by withdrawing from part of it, and this is pure Lurianic kabbalah. Before creating modern physics and calculus, Newton also spent years researching the proportions of the Temple in Jerusalem, which is a major concern of kabbalists, for they see the temple as a physical analogy to the many levels of the universe and the many levels of the human soul.

Newton was describing his relation to the Limitless when he wrote: "I do not know what I may appear to the world, but to myself I seem to have been only like a boy playing on the seashore, and diverting myself in now and then finding a smoother pebble or a prettier shell than ordinary, whilst the great ocean of truth lay all undiscovered before me." (Spence 462; quoted in Brewster 407)

From a logical point of view, talk of the Limitless seems to be, strictly speaking, meaningless. If it's truly beyond our thought, then we can't think about it. As Frank Ramsey said to Wittgenstein, if you can't say it, you can't say it and you can't whistle it either. And yet contra Ramsey, if at any point in the past we had said that reality was limited by the concepts we had at that time, we would have been wrong.

How should we resolve this contradiction? I think the clearest thing we can say of the Limitless is that it is the ideal limit of every possible conceptual revolution we or anybody else can ever have.

Suppose as a child you read a lot of romantic poetry—Blake, Wordsworth, Shelley—and became convinced that the world was alive and that all men were brothers. Then one day, you met an Ayn Randie, an apostle of the heartless novelist who believed the world was a ruthless machine, and the only honest attitude to take toward other people was as master to slave. If you took the Ayn Randie seriously, it would rock your world and open your mind to a completely different way of approaching people and reality. After reading a lot of books and having a lot of late-night debates in coffee shops, you might incorporate the Ayn Randie perspective into your world. And you would be absolutely unprepared for what it would be like to meet an honest-to-goodness cowboy, who thinks the two of you are just way too into city life and books and are ignoring what's important in life—the seasons, being outdoors with the cattle, knowing your gear, and being able to take care of yourself in hard weather. Meeting the Ayn Randie opened your mind, but it didn't prepare you for the opening of your mind from a totally different perspective that you got from meeting the cowboy. What's the next way your mind will be opened? Will you meet a refined and ironic prince from Ghana? Will you encounter a politician who knows exactly how to get a bill through the New York State Assembly? Will you commune with a sea turtle? You don't know. The universe is so unlimited, you don't even know where your limits are, or what it will be like when they are removed. The universe when all the limits are removed: that's the Limitless. It's inconceivable, which is just more Latin for you can't get a grip on it. It's not just hard to get a grip on it—it's in principle impossible, because to get a grasp on something means we have some way of controlling it, and we are always only a tiny infinitesimal portion within the Limitless. If you want to compare the Limitless to something, you should compare it to the universe itself, with everything in it, everything that will ever be in it, and everything that could possibly be within it—and all other universes that will ever come or could ever be.

That doesn't really clear things up, right? It doesn't help me relate to the Limitless to be told it is like itself. I would like a way to relate to or imagine or get a grip on or know the

Limitless using the brain and body that I have. And the Ari says okay, you can. You can compare it to three things in particular: a tree, a person, and a family.

The Ari loved trees. He invented new celebrations for Tu B'Shvat, the New Year for the trees, and named his work "The Tree of Life" after the tree in the Garden of Eden that promised everlasting life. If you look at Eleanor Davis's illustration of the Tree of Life on page 208, you can see why he thought a tree was a good image for the Limitless.

First of all, a tree is alive, so as living beings ourselves, we can "get it" on a nonconceptual level—we know about birth and flourishing and death because they're in our repertoire. Second, a tree has what is called fractal self-similarity: It is made of branches, each of which has smaller branches that look like the bigger ones. If you zoom in on any aspect of a tree, you will find greater and greater detail, like a coastline made of little inlets and peninsulas, each of which has its own inlets and peninsulas, and on many of the possible levels of description, this self-similarity continues ad infinitum. It's a system that is both unified and unlimited. Third, a tree has a certain dynamic interrelationship of parts and whole, so we can relate to a tree by unifying our left and right hemispheres—left hemisphere for details, right hemisphere for a "feeling for the organism." And a tree has a vertical axis and homeostatic mechanisms allowing matter and energy to circulate from top to bottom and bottom to top. This top-down vertical integration also characterizes

KETER
BINAH
CHOCHMAH
TIFERET
GEVURAH
CHESED
YESOD
HOD
NETZACH
MALCHUT

our experience of reality—which, as I have been arguing throughout this book, involves connecting the cognitive, the imaginative, the emotional, and the physical.

Why does this tree have ten circles on it? They are ten "spheres." A sphere is something like a particular way the Limitless contracts itself so we can relate to it; or looking at the same process from our point of view, it is a particular way we have of relating to the Limitless. The actual word is *"sefira,"* which is related to the English word "sapphire" and the Hebrew word for "number," but I'm translating it as "sphere" because of the mathematical suggestion, the evocation of "sphere of influence," and the fact that they sound kind of the same. If you look up at the sky, you see an endless expanse of blue. You can imagine that it's a giant sapphire. All the skies are limitless, but you can still see different skies, depending on the time of day and on your mood. That's what we should think of each sphere as—a limitless expanse, but a particular kind of limitless expanse. It's cognate with number because you can form infinite combinations of experience from the ten spheres, just as you can with the ten numbers. And just as the ten numbers reduce to two, one, and zero, the spheres reduce to two—an empty space and a ray of light shining into the empty space.

To get back to my story: I was looking for a way to connect to the world through thought, and a way to connect to other people through emotion, and a way to connect those two ways to connect. This is exactly what the Ari does with his spheres. Here's how it works. One sphere is called Chesed, which I am going to translate as "Giving." This is the way the universe presents itself to us when we experience it as overflowing with good: as a gift. Another sphere is Gevurah, which means judgment, strength, and limitation, which I am going to translate as "Drawing the Line." "Drawing the Line" refers to the limitations and delineations that make our experience of reality possible. When we try to understand reality, we experience some of the Limitless Giving—it blows our mind—and then we come up with some way to receive it, integrate it, and live with it—which is Drawing the Line. So, like the tree, we grow by alternating cycles of expanding and stabilizing, and with the right kind of self-analysis, you can take a core sample and count your rings too.

But the interplay of Giving and Drawing the Line also is a model of how we relate to another person. When we say of someone in a bad relationship that they have "no boundaries," we mean they don't have enough Drawing the Line. Every experience and every moment of our lives will have an aspect that overflows and an aspect that holds back so it can be the experience that it is. And in our relationships with other human beings, we are always going to be pushed and pulled between an aspect of the relationship that is open to greater possibilities and that forgives the other person for what came before, and an aspect of responsibility and accountability. Does that mean we have to always be schizophrenically whipsawing back and forth between responsibility and forgiveness, or ambivalently dithering about whether to love or to limit? No, the two of them can achieve a kind of dynamic optimal interrelationship according to the Ari, and that is another sphere, which he calls Tiferet, often translated as "beauty and splendor" but I am going to translate as "Give-and-Take." Just have in the back of your mind, if you can, that it's a Give-and-Take in which things work out beautifully.

This is a little different and a little the same as the annihilation of opposites William James experienced on nitrous.

It's true that like the world on whip-its, this is a system of opposites that eludes the grasp of thought. I can never *know* the correct balance of Giving and Drawing the Line. But we're not left with nothing to say or nothing to do, because I can take a position on the dynamic balance of the opposites within my own life. And the Ari recommends that for us, all ties should go to Giving. We acknowledge that the world is a mixture of love and limitation, but we would like more love, so we should live that way.

So even if we can't think about how to resolve the paradox, we can embody the resolution in our lives. Take for example how Jews tie their shoes. My great-grandfather was a forester for the emperor of Austro-Hungary, and I was always curious how that happened. Why did the emperor employ a Jew to take care of his forests? When I visited Safed to see the Ari's grave and take a bath in his ice-cold ritual bath, I had Friday-night dinner with some kabbalists and heard the story. Apparently a delegation of reform-minded Jews told the emperor that he should withdraw support from traditional Jews because they were obsessed with pointless rituals and meaningless laws, to the point of having rules about how to tie your shoes. This piqued the emperor's curiosity: If there were rules about how to tie your shoes, he wanted to know what they were. Possibly he also reasoned that if these Jews were so anal that they had a rule about how to tie their shoes, he could trust them to get other things done too. Whatever the emperor's reasoning, he gave them jobs in his government, including my great-grandfather's position as forester.

But what was the special way of tying shoes? It was a practical application or embodiment of the teaching of the Ari. Put on the right shoe first, because you want Giving to be paramount in your life, then put on the left shoe, which is Drawing the Line, then tie the left shoe so that Drawing the Line is limited and doesn't blight your life with limitation, then tie the right shoe, for more Giving, so you start your day with a Giving sandwich. Taking the system of spheres and embodying them in ritual with proper intention is a key part of the Ari's teachings. We want to relate to these concepts through the mind, the imagination, and the body, so they are cloaked in mental, imaginative, and ritualistic terms.

Like Lord Russell's theory of types, the Ari's spheres are an infinitely ramified system: There is Drawing the Line of Giving, and Giving of Drawing the Line, and Giving of Drawing the Line of Giving.

How does that work? Suppose my friend asks me for money

and I decide to give it to him. That is essentially a Giving action—I am allowing abundance to flow into his checking account. But is it also Drawing the Line? Am I making a separation? Sure—if I give him only what he asks me and not, say, all my money, then I'm limiting my flow of Giving with Drawing the Line. So that decision to give only what he asked me instead of all the money I have and the shirt off my back is Drawing the Line of Giving. But suppose on top of that, the reason I'm giving to him is I have a critical inner voice that is telling me that I should give more—that voice could be Drawing the Line of Drawing the Line of Giving. And I might feel that really the best way to show compassion is to limit my giving, because if I give too much, I will embarrass him—that might be Giving of Drawing the Line of Drawing the Line of Giving with maybe a little Give-and-Take thrown in. Complicated, right? In fact, infinitely so, because our experience can always go deeper and can always get more complicated.

Suppose Jane hurt Joe. She can ask, "Joe, do you forgive me?" once, and Joe can say yes once, which is an answer characteristic of Giving. But we all know that Joe might not really mean it, and we all know that Jane might know that Joe doesn't really mean it. So she can ask it again and he can answer again. There's no limit to the number of times, in theory, Jane can ask Joe to forgive her, and there is no limit to the number of times he can answer "yes, I do" in order to repair their relationship. In fact, you can prove there couldn't be a limit, because if Jane knew the limit was seven, and Joe forgave her seven times, she would have a good reason to wonder if he really did forgive her, because he answered yes only the conventional number of times.

All right, that's why the sphere of Giving is endless. But why is the way the different spheres relate also endless? Because Joe doesn't always have to say, in a Giving way, "Yes, I forgive you." Sometimes he might say, "I don't forgive you," which is a Drawing-the-Line response, and sometimes he might say, "Let's try to work it out," which is a response that's characteristic of the sphere of Give-and-Take. And any of these responses could hurt Jane's feelings, and Joe might have to ask her forgiveness, and of course Jane's response could be any of the three—I forgive you, let's work it out, or no. There are endless permutations of Drawing the Line, Giving, and Give-and-Take. If we keep our eyes peeled, we can see them everywhere.

However, if we try to grasp only one sphere, it will slip through our fingers. I have a friend who waited many years to have a child because he didn't want to shame his two childless older brothers. My friend was afraid that if he pursued his own happiness by having a child, he would be condemning the life choices of his brothers. So you could say that he was acting from an abundance of Giving, doing everything he could to avoid using Drawing the Line against his brothers. But by avoiding judging his brothers, he was in fact being very judgmental to himself—he was judging his innocent desire for happiness as being overly judgmental! So his attempt to be purely Givingesque slipped over into its opposite in a big way. We can see this happen in a larger arena as well. Cultural relativism comes from a kindly desire to avoid the arrogance of colonialism, but if you're a woman who was caught necking, and your cousins are getting ready to stone you, the local Westerner's kindly desire to refrain from judgment can feel pretty harsh.

The Ari's picture of knowledge is, I think, a clear improvement on the correspondence theory of truth we talked about when we were looking at the apple in the head and the world's largest game of red rover. The correspondence theory puts us outside the universe, trying to form a coherent picture of it in our mind. If the picture is incoherent, it means that our ideas are not clear, and we need to clarify them. For the Ari, by contrast, we are part of a growing tree that is pushing out in all directions. As we think harder, we push the tree out farther, so as a consequence our thinking doesn't reduce the number of contradictions in our lives, it increases them. However, if we play our cards right and live in an integrated manner, the result won't be incoherence or confusion, it will be an ability to flourish in more situations and interact fruitfully with more kinds of people. If the Ari is correct, then we understand why we experience so many contradictions. Of course we can't find a resting point between intuition and reason; of course we are not sure if we should show love and risk being taken advantage of, or hold back and risk severing our connections; of course we don't know if we should be humble or if that means we are proud of our humility and therefore not humble; and of course we don't have a single answer to the question "Should we engage with reality, or feel, or think about it?" The Ari's general approach lets us understand why the more questions we can ask but can't answer, and the more paradoxes we understand but can't resolve, the better off we are. We are part of a bigger, more flourishing tree of life. Paradox is not a sign of a problem, it's a sign of success.

14
Faces of What

Logician and Taoist Raymond Smullyan tells a good story about the Buddha. Once the Buddha came to town and gave everybody the chance to ask one and only one question. One man really didn't want to blow his big chance, so he thought and thought and finally asked, "What is the best question for me to ask and what is the answer?" The Buddha said: "That was, and this is the answer."

I made up a riddle inspired by this story:

Q: It usually starts a question but now it ends an answer. What is it?

A: What.

That is, the word "what" usually starts a question, but here—in this riddle—it ends the answer.

My favorite Nazi philosopher, Martin Heidegger, once defined "human being" as that being for which the question "What is it?" is an issue. He calls human beings Dasein—German for "there being"—but he might as well have called us "where being?" because we're like the empty space of the Ari—we get to know ourselves better, not by adding definitions or learning facts but by asking more and deeper questions. We don't just happen to have the question "What am I?" Our defining feature is that we're the thing that can ask what it is.

For the Ari, wisdom is called *chochmah,* which kabbalists explain means *koach ma*—the power of "What?" So now that we've gone through spheres and the Limitless, and the contraction of the Limitless, and the sad and happy story of my life, which, now that you've read my book, is the sad and happy story of *your* life, we should ask the big question: "So what?"

What are we supposed to do with the Limitless? Or, put another way, what are we supposed to do with our lives, which are all about wrestling with limitations, sometimes overcoming them and sometimes getting beaten up by them, if reality is actually limitless? Why does it matter? How are we supposed to think about the limitations on our minds? In psychology we are often learning about limitations on cognition—little modules of our minds or brains that cause us to see optical illusions or misunderstand an interval of time or hear two sounds as similar that are really different. Or we look at animals, which are limited versions of humans, or mentally ill people, who are limited versions of mentally healthy people. What is the limit of all this talk about limits? When do we actually get to know what really is and what we really are?

We're faced with the question "What?" What exists and what does it matter? If you look at that simplest form of that sentence—"We're faced with 'What?'"—you can see that there are three ways into it. The first is through thinking about who *we* are, and I've been doing that in this section—telling you a little bit about my own personal craziness so you can see what you relate to and what you don't, and we can form a "we." The second part, the "What?," has been what we've been investigating with our philosophy and science: the question of what is. The Ari's approach focuses on a third way in: the facing. Since the question "What?" faces us, he wants to talk about its face.

Are we allowed to? If we think of the Limitless as having a face, aren't we violating both the hygiene of modern thought and the ancient Jewish commandment against idolatry? Aren't we charging backward in time and going back to the notion of an invisible superman in the sky? Haven't we just given up on modernity and embraced myth?

Well, look at the word "embrace." We have minds that conceive and we have arms that embrace; "conceive" sounds mental and "embrace" sounds physical. It's really no more or less human to say we have arms that can embrace reality than to say we have brains that can conceive it. That's why the ten spheres are drawn in the shape of a person: for example, Giving and Drawing the Line are his left and right arms, which we use in hugging another person's body close, and Give-and-Take is the heart. "Conceive" is just Latin for "grab together" anyway. Since the Limitless is limitless, it's just as infinitely far from our

cognitive abilities as it is from our arms and legs. Or looked at another way, if the Limitless is really limitless, it ought to be able to limit itself: If it couldn't, that would be a limitation.

But faces? Look at Ms. Davis's picture of the tree.

You can see a face in the sky and a face in the trunk of the tree. Are the faces an illusion and the tree real? No, the tree's not real either—there's just ink on paper. But those two faces are there just as much as the tree for sure.

The Ari calls the humanlike aspects that the Limitless presents to human consciousness Faces. This is his primary innovation compared to older kabbalists, and it's what makes his teaching so risky and scandalous, both to Enlightenment thinkers and to traditional Jews, who, as Weber pointed out, have more in common than meets the eye. For the thinkers of the Enlightenment, if we say reality has a face, we are just projecting our infantile needs and desires on it. For traditional Jews, if we say God has a face, we are violating that second commandment. Kabbalists have had to be very careful, therefore, to explain just in what sense and how it is appropriate to relate to our ideas about God as real. Here is an Italian interpreter of the Ari who lived his life one step ahead of rabbinic excommunication, Rabbi Moshe Chaim Luzzatto, arguing that the Ari is not idolatrous:

> The [spheres] can shine with an abundant radiation of light or with diminished radiation. They can appear in numerous different forms and likenesses, although in truth they possess no form or appearance. It is just that they

> appear in some form or likeness, but one who examines them will see that in truth, the form and likeness are purely contingent upon the observer, as stated in the verse, "And in the hand of the prophets I have used likenesses." (Hosea 12:11) In their essence, however, the [spheres] are only an extended array of powers organized in their necessary order, interdependent in their various laws and influenced by one another, as required according to the perfection of the entire ordered plan. (Luzzatto 25)

Luzzatto is arguing above that the images that appear in our minds when we try to understand the Limitless are not what it actually looks like, because it doesn't look like anything. In itself it's unlimited, and "how something looks" is a limitation—it refers to how it interacts with a being with a particular kind of mind and a particular kind of visual system.

Neuroscience has just started to explain why sometimes we respond to faces and sometimes we don't. In *The Polyvagal Theory*, Stephen Porges gives an account of how we are able to respond to faces along the way of advancing a new theory of the parasympathetic nervous system that lets him address what he calls the vagal paradox. The big picture in neuroscience is that our sophisticated cerebral cortex is a relatively recent development that sits on an ancient nervous system that goes all the way down to our butts. One of the biggest brain-to-butt wires is the vagus nerve, which connects the cranium to the face, the larynx, the heart, and the guts. The vagal paradox is that the

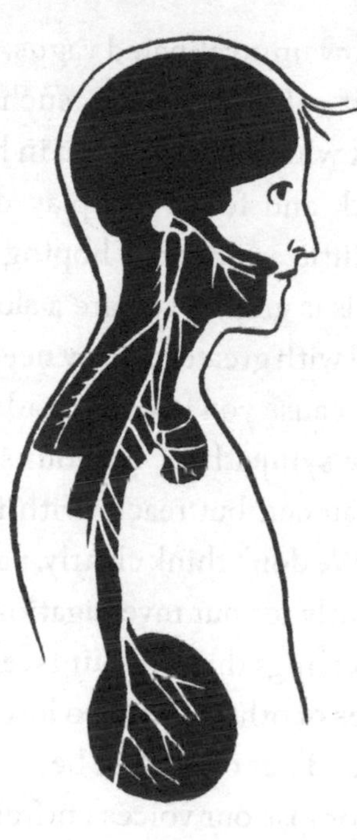

same vagus nerve that causes the slowing of the heart during "rest and digest" behaviors also causes the potentially lethal drop in heart rate called bradycardia. The answer provided by the polyvagal theory is that there are two different vagal pathways from the brain stem to the organs: a faster one in which the axons of the nerve cells are insulated with a white fatty substance called myelin (the myelinated vagus) and a slower one where they aren't (the unmyelinated vagus).

If we add the sympathetic nervous system, that gives us three neural pathways, each of which supports a different way of reacting to the life.

The first, the slow unmyelinated vagus, takes over when we are in situations of extreme trauma, such as when the forty-mile-an-hour great white shark has you in his jaws and is whipping his head back and forth. We play dead and lower our metabolism to as little as possible, hoping the dangerous situation will pass. This is great if you are a slow rumbling reptile; if you're a mammal with greater energy needs, this sudden drop in metabolism can cause you to drop dead.

The second, the sympathetic nervous system, takes charge when we are threatened but reacts with fight-or-flight rather than a shutdown. We don't think clearly, past and future tunnel out, and, significantly for our investigation, we lose the ability to express subtle feelings through our face and to detect subtle feelings in the faces of others. We also lose the ability to detect the human voice, and our own voice becomes less human—we screech or squeal or lose our voices entirely. The art of politics sometimes seems to be largely a matter of getting people to think with their sympathetic nervous systems.

The third neural pathway, the myelinated vagus, puts a "vagal brake" on the sympathetic nervous system and its simplistic fight-or-flight approach to reality. In this mode of being-in-the-world, we both respond to modulations of facial expression and voice and have well-modulated voices and expressive faces ourselves. We play. We laugh. We communicate. The myelinated vagus is responsible for our ability to suck, which is interesting since, according to Isaac the Blind, an early kabbalist who lived about a hundred years before the Ari, we relate to the highest spheres not by knowing but by sucking. The myelinated vagus also pulls on the bones of the inner ear so that loud rumbling sounds are quieted and high-pitched sounds are amplified. This is because when we were little shrewlike mammals, we evolved a secret radio frequency that the dinosaurs couldn't hear.

> The evolution of the mammalian middle ear enabled mammals to communicate in a frequency band that could not be detected by reptiles that, due to a dependence on bone conduction, were able to hear predominantly lower frequencies. The ability to hear low-amplitude, high-frequency airborne sounds is accomplished when the middle ear muscles are tensed to create rigidity along the ossicular chain. This mechanism unmasks the high-frequency sounds associated with mammalian vocalizations from background sounds. (Porges 206)

These days we can automatically switch the secret mammal radio on or off by pulling on the bones of the middle ear:

> During social interactions the stiffening of the ossicular chain actively changes the transfer function of the middle ear, and functionally dampens low-frequency sounds and improves the ability to extract conspecific vocalizations. However, the selectivity to conspecific vocalizations comes at a cost, and the detection of lower acoustic frequencies generated by predators becomes more difficult. (Porges 214)

We use the myelinated vagus to tune in on talk when we're safe in our houses or burrows but not when we're scared out of our wits by dinosaurs. When the myelinated vagus is online, we hear voices and see faces; when it's not, we don't.

So in the light of contemporary neuroscience, when the Ari says that the Limitless has faces, he means that it is okay to relate to the Limitless with our myelinated vagus. When we say

that God has a face, we don't mean that we're going to look out the window and see a giant face looking at us.

That would be weird. We mean that the whole of reality, the texture of our lives, can be related to in that third way—a way in which we respond to subtle, warm emotional ebbs and flows of meaning. We hear the voice of God, but we do not parse the words of God. We feel a presence, as we do when we sense without meeting the owner that a house belongs to somebody. And we feel like the universe is our home. Or burrow. Or our parents' closet.

In my view, these parallels between contemporary neuroscience and sixteenth-century Jewish esotericism are certainly less than a corroboration of either but a bit more than just an analogy. I think the parallels occur because both are attempts by human beings to understand themselves and both avoid the mistakes of Descartes. Both view the human being as fundamentally unified, both see knowledge as the evolution of life, and both view self-knowledge as a stage in this evolution seen from the inside. The Ari is providing a phenomenology of the growth of consciousness, and neuroscience is confirming the existence of certain structural features, corroborating that we have a right and left side (the two hemispheres), for example, and higher and lower feedback loops for interacting with the environment.

An atheist and, for all I know, Porges himself might respond: "Maybe the myelinated vagus will cause us to relate to reality as we relate to a human face, but that is a delusion. We are taking the part of our brain that is able to detect emotional meaning where it actually exists and applying it to things where it does not exist. It is a mistake." But Isaac Luria can answer: "Why is it any more appropriate to relate to the world through the cerebral cortex than through the limbic system? If seeing reality as having a face is a product of the human brain, then so is seeing reality as being only a source of danger and safety, or as a play of ones and zeroes, or as an array of three-dimensional objects in time and space. How do we know that it is inappropriate to respond to the whole Limitless as playing with us, loving us, laughing with us, crying with us, teasing us, winking at us, smiling at us, sometimes frowning at us, and talking to us?"

Absent a brain and eye, there is no way the world looks, and absent a brain and ear, there is no way it sounds. Once we have a brain, it looks and sounds different ways, depending on how we use that brain. If we are in a state of terror, we cannot respond to the faces and voices of other human beings. If we are cruel to animals, as the dog dissector Descartes was, we cannot respond to the faces of animals. If we are shut down in terror, we certainly cannot respond to the faces of animals or human beings. People who suffer from autism have trouble responding to human faces. The Ari is arguing that materialist atheists are like autism sufferers when it comes to the faces of the Limitless. They close themselves off to the moods and hints of the Limitless as it manifests in their lives.

If we disagree that the Limitless is like a person, I think we need to ask ourselves: What is it to be like a person? Are we like a person? Is our dog like a person? Suppose I am looking for sex, but I view it as a physical release. Then I am viewing a person—myself—as nothing more mysterious than a gasket that builds up pressure and then needs a release. If that's my view of what it is to be a person, then why shouldn't I believe the universe is anthropomorphic—that is, that its "morphos" (form) is like "anthropos" (a man)? All I mean by "anthropos" is a mechanical system. Or suppose I view myself as fundamentally an animal. In that case, if I view the Limitless anthropomorphically, I'm just saying it is like an animal. Is the Limitless like an animal? Well, animal life is constantly exploring the environment and evolving, so of course it's *like* an animal inasmuch as the best grasp we can get of an ideal limit is through the process of trying to reach it. If I'm a romantic and I view another person as a mystery, then to say the Limitless is like a person is to say it is a mystery. If I view my partner as an independent source of meaning and mattering whom I cocreate reality with, then personalizing the Limitless is just to believe that I can cocreate reality with it too.

Now, at this point somebody—you, maybe, or me when I'm not feeling it—may want to deny that reality wants anything or that we can have a relationship with reality, or life, or the universe in any way that is deeper than simply projecting our own wishes and fantasies on an array of unfeeling stuff. That is a way of saying that it is reasonable to respond to reality with a fight-or-flight response but not reasonable to react to it with the responses that typify the myelinated vagus—communication, playfulness, listening, and facial perception. Reality doesn't have a face. People have faces.

But do they? People have muscles and skin and holes for light and air and food to get in. But they have faces only for us when we look for faces and interact with faces. In the same way, the faces of the Limitless are not there other than as they interact with our contingent human consciousness. You might want to draw the line at talk about the Limitless wanting anything from us and say the Limitless didn't have a plan—it just is and we came about randomly. But why? To call the Limitless random rather than planned is to limit it. Randomness and order are both constructs of evolving human consciousness.

So does God exist? One way of describing belief in God is saying that, if you make a list of all the objects in the world—Volvos, cheeseburgers, mountains, quarks, neutron stars—one of the things on the list will be a superbeing who created it all, named God. The Ari does not believe that, because that would be to limit God, and he believes the single most accurate thing you can say about God is that He is limitless. The Ari says God does not belong on the list of things that exist, so in a sense, he is closer to an atheist than to some believers. Another way of describing belief in God is that there are some things it is appropriate to respond to with the emotional responses that typify the myelinated vagus. They include human beings, animals, and our experience of life as a whole. The Ari thinks that that is right.

If the Limitless presents a face to us, what does that mean for contradiction? The Ari is interested in two particular faces, which he calls Arikh Anpin and Zer Anpin—literally translated as "The Long Face" and "The Short Face," but here "long" means patient as in "long-suffering," so I will translate them as the Gentle Face and the Challenging Face. The Ari uses these two faces to explain the problem of evil: How could an infinitely powerful and good being create the world where the *New York Post* has so many shockers to write about? You feel like saying to this being, "How could you?" in both senses of the phrase.

Luzzatto, the Ari's Italian disciple, argues that if the world were set up so we just got pleasure, we would feel a little gross, a little ashamed. We want to feel unlimited, and if we are just taking happiness as a handout from an outside source, he believes we would chafe at the limitation. So there is a sense in which we want life to be the sort of thing that we can succeed and fail at, and therefore, unfortunately, it has to be the sort of thing that we can royally screw up.

Marcel Proust brings down the following insight:

> We have to rediscover, to re-apprehend, to make ourselves fully aware of that reality, remote from our daily preoccupations, from which we separate ourselves by an ever greater gulf as the conventional knowledge which we substitute for it grows thicker and more impermeable, that reality which it is very easy for us to die without ever having known and which is, quite simply, our life. (Proust 298)

Proust is reminding us that by living our lives, we are running a risk—that we can be so absorbed in triviality that we die without ever having known our own life. Henry James in "The Beast in the Jungle" warns us that triviality isn't the only way to lose our lives. His protagonist, Mr. Marcher, is so sure that at some point in his life his unique extraordinary destiny will reveal itself to him that he never notices his best friend has been in love with him her entire life, and when he realizes it, it is too late. She's dead and he understands that the special destiny in store for him was to be the man who most completely missed the point of his life.

Whenever we are worried that our life might be the kind of thing such that we can miss it or waste it, we are experiencing the Limitless presenting itself to us as the Challenging Face. It's not that there is a superbeing that is angry at us, but we are measuring our life in such a way that we could fail. The Challenging Face is angry in the sense that we can let it down—we can meet the challenge or fail to meet it. When we're encountering the Challenging Face, we feel like the children of a short-tempered overworked dad or mom, who wants us to do well on our math homework, who wants us not to be a jerk to our little sister, and who is disappointed or even angry at us if we let him or her down. We don't have to imagine there is a literal giant human being outside who is going to squash us like a bug—as long as there is the conceptual possibility of not living our lives as best we can, we are face-to-face with the Challenging Face. The sense that the world needs healing, and we evade that at the peril of our souls, is the sense of the Challenging Face.

But there is another aspect of life where we don't feel we need to add anything at all to the smell of grass or a lungful of air, or the chance to look at a person. Pure existence is enough to be grateful for. We don't need to make it significant or succeed at it or pass a test. That's the Gentle Face. We are having an experience of the Gentle Face whenever we're grateful for the opportunity to be us, to exist, and to have a world at all.

What's the relationship between the Gentle and the Challenging Face? The same as the relationship between your angry and your happy face—they are two different expressions of a single reality. They are not two guys. The Ari denies that reality at its deepest is split, because splitting would be a limitation. There are not, as the gnostics thought, "two powers in heaven." Mommy is not a good breast and a bad breast. There is not a God in heaven and Satan down on Earth. There is not even a God creator and a matter that he works on. Life's challenging moments and its gentle moments are just different facial expressions—different miens or aspects or moods that the Limitless (as it relates to human consciousness) can be in.

But given that, the Gentle Face is closer to the reality of the Limitless than the Challenging Face. What does that mean? How can a face be closer to an underlying reality if that underlying reality is limitless? Because, the Ari argues (or perceives [or imagines]), the ultimate reality wants us to experience limitlessness and joy. The judgment of the Challenging Face is just a means to an end, which is to give us moments of limitless joy and growth. When we tell our children we want something from them—we want them to succeed and turn themselves into good people—we are presenting the Challenging Face to them. When we let it slip that we would love them whatever they did, we are revealing the Gentle Face, which is more real, closer to what our real attitude is.

15
The Point of It All

When we started together, I said that if you happen not to believe in Santa Claus, you should take whatever you think the point of your life is, given that you're going to die, and treat that as your personal Santa.

The question about the point of life is really two questions: "What is the point of life?" and "How could life have a point?" The first question is easier. The point of your life is probably what you think the point of your life is—to fix what needs fixing. If the Mafia is taking over your neighborhood, you've got to do something about that, and if you're smart enough to cure Alzheimer's disease, kindly get on that. Both projects—Mafia fighting and Alzheimer's disease curing—and innumerable others present themselves to you, and all of them can be your life's point. The harder question is "How could our lives be the kind of thing that has a point? Why would it matter? Why is there a world that needs fixing? Why are there Mafias and dementias and baby-eating snakes?" The Ari's answer to the question of why the Limitless would manifest itself as a broken, wounded world full of suffering is that that was the only way to give our lives a point, and without a point for our lives we would really suffer. By overcoming the imperfections of the world, and cocreating heaven on Earth, we get to be important rather than ashamed.

In a sense, then, as Newton told Locke, the Limitless contracts itself and creates a space of apparent emptiness in which we can operate. I say "apparent" because the contraction of the Limitless is a deeply paradoxical concept. The Limitless has to be able to withdraw itself in order to create limitation, but even when it withdraws itself, it has to still be there. If the Limitless can't withdraw and give us some room to operate, it isn't limitless. But if the Limitless can withdraw and really not be there, it isn't limitless either. It can and it can't. This, as Aryeh Kaplan pointed out, is the source of the famous puzzle of whether God can make a rock so heavy he can't lift it. (Kaplan, "Paradoxes," 131–39) Its relationship to us is a little like the relationship of husband to wife in "The Gift of the Magi"—it can't think too much about the gift of freedom it wants to give us or the gift goes away. A point for our lives is the best gift the Limitless can give us, and to give it to us, it has to give us some room.

Is this answer too cute? Or specious? Or illogical? It might be all three, but to be fair, we have to compare it to other attempts to answer the question "How can life have a point?" There are a few different tacks people can take.

One tack is to reject the question. "I get a job to make money to have kids because *that's just what you do*. Back off. What's your problem?" That is an attitude that takes some people through their whole lives and many people through much of their lives. If you are satisfied with it, there isn't much anybody can tell you. You are in a sense living the life of an animal. If the dog could talk and we asked him why he eats out of the cat's litter box, he'd say something similar: "That's just what you do. Why are you trying to harsh my buzz?" It works out okay for the animals more or less. I mean, not if somebody eats them or peels off their skin to make shoes, but on a day-to-day level, some animals do okay. I'm not knocking animals.

A second answer is that the question "What is the point of doing anything?" doesn't make any sense. This is in the spirit of the logical positivists whom we met earlier. You can ask, "What's the point of wearing a sweater on a cold day?" but you can't ask, "Why should I do anything?" I think this is hard to argue with, and entirely impossible to believe.

A third answer is that there is a superbeing of some kind who dictates what we should do, and the answer to all "why" questions is "Because he wants it." I've always found this just impossible to believe, perhaps because it implies that some of the most interesting and unique features of human beings—freedom, judgment, and the capacity for criticism—play no role in the ultimate scheme.

A fourth answer is that life has a point just because we human beings choose to give it one. So the universe has two kinds of things in it—matter, which is intrinsically pointless, and human beings, who have the ability to freely bestow a point on something. This is Descartes's eagle-shark or mind-body dichotomy all over again and raises the same problem: How is it possible there could be such beings? How do point bestowers manage to exist in a world of pointlessness? And if there were such beings who are able to create purpose, does anything keep them from changing the purpose? If they can lay down the law in one second, can't they pick it up the next second? And if they can, what meaning is there in their laying down the law to begin with?

So if you think this kind of question needs an answer (and I admit I can't prove to you that it does), I think the Ari's answer can lay claim to being a good one.

The Limitless analogically can be said to want something—namely to bestow its limitlessness.

It bestows its limitlessness on us by creating a physical universe that ultimately evolves to the point of creatures who are able to program themselves, freeing themselves from biological constraint and cultural conditioning.

These beings (us) have a need to be limitless, but their limitlessness also requires them to exist in a situation in which things matter. Hence an unfinished, unhealed universe, shot through with beauty and horror, but one that each of us is able to repair in a unique way.

And that's where we are. As we remove limitations and remove suffering from each other and are able to grow into beings who have more and deeper experiences and discoveries, we become aware of ourselves as expressions of the unlimited. And then the Limitless—manifesting as all of us—continues to be unlimited. And that's both the point of everything and the explanation of how it could be that everything, including us, has a point.

Why is it so confusing? Why did it take such a long journey to get here? I think because just as we need to fix the world by putting pieces together, we need to fix our understanding of the world by putting pieces together, and each of us has different pieces.

We have looked at how the Ari compares the Limitless to a tree and to a person: The final thing he compares the Limitless to is a human family. So just as a husband and a wife have different ways of looking at the world but are able to blend their opposites in a productive, procreative way, so the Limitless displays itself as finite by taking the form of different people with opposing points of view who are still able to get along and love each other and form families. They also fight and kill each other, but every war, every genocide in human history is a Thanksgiving Day dinner argument writ large. We are all family members and all blood shed is brother against brother. We also embody the Limitless in all the ways we relate to each other through genealogical time. We participate in the sibling-to-sibling relationship, the child-to-parent relationship, the child-to-grandparent relationship, and the child-to-great-grandparent relationship. The degree to which we are able and willing to forgive injury depends on how close we are in the generational map. If we are brothers, we don't forgive much in the way of error, and demand perfection. As parents, we want some perfection from our children, but we also expect there to be diapers that need changing. As grandparents, we are not so interested in forcing our grandchildren through the grueling tasks of life—we just give them cookies. So the ultimate home of paradox lies in the fact that the members of this ramified human family love and hate each other and take all imaginable positions on the problems we all face.

You might believe it or you might not. I think I believe it, at least on good days. More importantly, on the way to believing it, I resolved the question that I started with. Because the Gentle Face—the aspect of the Limitless that loves us no matter what, that bestows blessings and gifts at every moment, looks a little familiar. The Gentle Face is described in the kabbalah as having a long white beard. Now, who does that remind you of, giving gifts regardless of whether we are deserving and sporting a long white beard and a kindly face?

Santa Claus.

Santa Claus is a way of imagining the aspect of the Limitless that the Ari imagines as the Gentle Face—the long compassionate face with a long beard down to his chest, white as snow, connecting his brain to his heart.

Does Santa Claus exist? Yes. He is a face the Limitless takes on when manifesting itself to human consciousness.

Is Santa Claus also your parents?

Why not.

But that is not quite the last word of this investigation.

Certainly it's suspicious that Eric Kaplan, a philosophically trained, Buddhist-ordained, Jewish comedy writer is putting forward a theory of reality that is a mash-up of philosophy, mysticism, comedy, and Judaism. And what's worse, it seems that I have abandoned one of the most popular features of Judaism, its prohibition against proselytizing, by telling people that Santa is actually a concept from a sixteenth-century kabbalist from Safed. Are we back to "Your doxy is heterodoxy, my doxy is orthodoxy"?

I'm not really advocating that you take my mixture and swallow it. I'm just showing you how I did it so you can do it yourself. So my brain and my life have gotten to a place where I can more or less function by putting together these reflections on logic, mysticism, comedy, and the kabbalah. If you're a half-Belgian, half-Pakistani lion-tamer atheist, I would expect your take on Santa Claus would include a mixture of Sufism, atheism, lion lore, and Luc Sante. I'm not saying I'm not trying to start a cult—of course I am. Cult leader is a great job, and anybody who had the chance to apply would be crazy not to. But you should start a cult too, and I'll let you be in my cult if I can be in yours.

Whether you join my cult or start yours, I do think that it is healthy and sane to look at questions of what exists or what is true in the context of what kind of life we would like to live, both as individuals and as a community.

Up until now, when we've asked the question "Does Santa exist?" or "Do I want to be a logician or a mystic?" we have projected ourselves into the future and into other possible states of being. Eddie, the brain helmet that makes you believe in Santa, Neurath's boat—they're all ways of imagining belief as a trip into the future, and we've tried different ways to answer the question of exactly what ticket we'd like to buy from that world-historical travel agent. But all these questions about future selves beg a more fundamental question, which is how we want to experience life right now. And luckily, that ought to be the easiest question of all, because to answer it we don't need to worry about philosophy or science. You can answer that question by just becoming mindful of this moment and how you feel about it—emotionally, intellectually, and physically. When you do, you will be able to bring your whole being on line to come up with a response, not just your mind.

Let's try.

What do you want from each moment of life? If this moment were going to last forever, what would you like it to be? What's on your shopping list?

Well, one thing you wouldn't want is to be absentminded during it, because if you were, you would miss your only moment! So you would want a feeling of presence.

What else?

Would you want to experience an itch? No. Nausea? No. Physical pain? No, obviously that would be awful.

You'd probably want pleasure. Neuroscience tells us there

are two kinds of pleasure: the pleasure of thinking that if you play your cards right, you will get an orgasm (or food)—which comes from dopamine—and the pleasure of satisfaction itself—which comes from endorphins. I think to have the feeling of being about to have pleasure for infinity but never getting it would be a tremendous rip-off. I wouldn't wish that on my worst enemy, and so far, my relationship with you has been nice. So let's go with the second one.

So first attempt—if you could have one moment left, either because it will be prolonged infinitely or because you will be snuffed out after it (I hope not the second, because I like you)—you would like to be an orgasm or an endorphin high.

Or would you? I admit it beats a kick in the teeth. It definitely gets partial credit! But most people who have the opportunity of directly administering endorphin highs to themselves by shooting up heroin, or taking Oxycodone, the hillbilly heroin, choose not to. For example, me. I took Oxycodone when I needed to donate bone marrow for my brother, and it was ridiculous how happy it made me feel. I was literally singing a happy song in a cancer hospital. It was also humiliating to learn I had a switch in my head that could be turned from SAD to HAPPY by some stupid chemical, and I was very glad when I ran out of Oxy! People who do end up picking a self-administered endorphin high—opiate and heroin addicts—aren't exactly happy with their life choice. It's degrading and brutalizing, a form of slavery.

But in this thought experiment, you are allowed to pick

whatever our moment would be. Could you take the orgasm and add some sprinkles on top that would make it worthwhile as the only moment you're ever going to get? The actual moment of pleasure is, I think, pretty difficult to remember because in a sense you were not there. So if you pick your only remaining moment of life to be an orgasm, in a sense that is suicide. You won't be there to enjoy it. And as I mentioned, I don't want that for you. I like you. Big fan.

So let's add awareness. But what would you like to be aware of? You don't want to be aware of something distracting, obviously. It's your only remaining moment, so I think you want to be present for it. So you should be aware of something good.

I think this is starting to sound pretty good. Why? Why does it solve the problem of the heroin addict? Because your best moments don't just feel a certain way—they actually accomplish something that you care about. They matter. When we feel trivial and disconnected, it feels bad.

I hope you can now appreciate why it was such a terrific idea to put the question of Santa Claus and existence in general in the context of your own life, because now you can approach the problem of your favorite moment with your whole being, and not just intellectually. Because we're running at it with your whole self, you can access the wisdom of bodily sensation. And your whole self tells you, if you're anything like me, that narcissism and solipsistic enjoyment make you feel bad: The idea gives one a kind of queasy, seasick feeling. Our gut read on the situation is that if we are floating brains in the Matrix, experiencing incredible pleasure, it feels weird and creepy. I think we feel that it's both not good and certainly not the best moment you're capable of. In other words, although there may be a selfish, disconnected mood in which you say, given the choice between the following two moments:

(a) experiencing incredible pleasure while making the world suffer

and

(b) experiencing a little bit of pain that makes the world feel okay,

you would pick (a).

There is no normal, healthy mood in which, given the choice between

(a) experiencing incredible pleasure that makes the world suffer

and

(b) experiencing incredible pleasure that makes the world feel good,

you would pick (a).

I know you, and you're not that kind of person! As I said, big fan. So for you (I don't know about your sister-in-law, but for *you*) if it's your last moment, all things being equal, you don't want to spend it being spiteful. You would like it to be a good one not just for you but for others. It just *feels* better that way.

So what if we craft a situation in which your pleasure is super important? What if the universe is in danger from a monster who is about to eat it, and all that's needed to defeat him is for you to have an orgasm? That way in your final moment, you will get to be the greatest hero who ever lived and have the greatest pleasure anybody ever had. That's better than the Oxy-addict moment except it involves conjuring into existence a universe-eating monster. Scary! Why do you need to create a universe-eating monster to have a nice moment? And once you've created the poor hungry guy, why do you have to kill him? Not to be all judgey or anything, but who's the real monster here?

Okay. What if the orgasm was doing something really good that did not require killing or monsters?

What if there were another person, and your orgasm gave that person an equally great orgasm?

Now we're getting somewhere! Can it get even better? Sure. What if it taught that person to have an orgasm for the very first time—they were sexually blocked or a virgin and it was only your lovemaking that gave them bliss? I like this! What if instead of it being one person, it was every person—an infinite number of people? That is starting to sound very good.

But I'd like to raise a problem. A lot of you are going to say, "Problem? Infinite orgasm for everybody, I don't see a problem. I don't *want* to see a problem. I'm buying it, moving on." And that's the problem. It's boring and it's obvious. It sounds like some kind of dumb *Penthouse* Forum meets *Wired* magazine double issue. Probably because it is laid down by our biology in a sense, but also because we get that message from our media and our advertising 24/7. And there's nothing wrong with doing something fun again because it was fun the first time, but it's not the best. All other things being equal, the best moments are fresh and surprising. You don't want to be bored.

What could you add to our moment, considering that it's already super pleasurable for us and causes super pleasure in infinite other people? How can it be made not boring and not a cliché?

Some great pleasures that are not boring come from discoveries—learning something new and solving problems. They can be intellectual or artistic or personal—the moment when you figure out what your calling is, or what kind of pet to get, or whom to marry, or how to build a house. The experience of coming up with a great new solution can't be beat. So what if your perfect moment was one when you discovered some amazing new way to make yourself and everybody else feel amazing? And it was so different and so amazing that, compared to it, the moment before was like the evil monster swallowing the universe. It was such an "aha!" moment that it made what happened before seem dull by comparison. It was like the feeling Newton had when he saw his apple drop, or the feeling Beethoven had when he thought "bum-bum-bum-BUM!"

That's a mental analogy. We had a physical analogy. How

about an emotional one? Emotionally, as the song "Nature Boy" goes, the greatest thing is to learn that we're loved and to love in return.

So . . . the best moment would be one that gives you maximum physical pleasure, emotional satisfaction like learning you love and you're loved in return, mental pleasure like the "aha!" moment, and all in a way that's so fresh and exciting, it makes your life up to that moment seem like the fifth cup you made with the same tea bag.

That's pretty good, but not perfect. I don't want you to feel bad about the moment before!

What kind of relation do I want you to have with the moment before? I think you would like the present moment to be different from the moment before and better than it, but not different and better in a way that makes the moment before look, in retrospect, like garbage. You want to be grateful for the previous moment but still enjoy a present moment that is inconceivably greater, richer, fresher, more interesting, and more enjoyable.

In other words, the best moment doesn't just throw away the old moment and replace it with something new. The best moment grows out of the one before like a strong creative adult grows out of a fun baby, or a big green broccoli plant grows out of a two-leaf seedling.

Now, what if you think growth is boring? What if you think the way life produces a full-grown broccoli out of a two-leaf plant, or a creative virile or fertile adult out of a little baby, or the whole biodiversity of Earth out of bacteria is boring? Then sadly, you're out of your mind. If you prefer death to life, then I really do think you need help. Stop reading this book and throw yourself into life. Grow a broccoli. Take care of a child. Milk a goat. Walk. Notice the cycles of your own body. This is my one dogmatic claim and I'm sticking by it with the tenacity and self-righteousness of a Super-Pope: If you're bored with life, the problem is with you, not life, because life is not boring. Plants, animals, ecosystems are the most interesting things there are. The proper response to boredom with life is to get into the body, get into the emotions, and get into the pulse of other lives—human or nonhuman.

So you are looking for a moment that matters, one that is fresh and limitless and that grows out of the previous moment organically like a shoot pushing up from the dirt. My claim is that Santa Claus or a point to life exists if believing in them puts you in a brain state that can give you a moment like that. And the only way to know if you are in that sort of state right now, or if you are getting closer to it or further away, is by integrating your brain right now and seeing if you are. But couldn't you know if it mattered without going through it because a really smart scientist told you, or a great priest or a mystic, or Wikipedia? No. The most interesting things that matter happen because you learn how and why they matter by doing them. Yes, there are a lot of things that matter and you knew they would matter before you did them. Like making my kid learn Spanish probably matters, and I knew that when I was socialized into being a middle-class parent and had "Make your kids learn things: It matters!" drummed into my skull. But that's the more limiting, and boring, kind of mattering. The best kind of mattering is when this moment for you is a gateway into realizing that a whole bunch of things matter that you didn't know before and in a way you didn't know before. And, by definition, you can't know that before you live through the moment.

So, if you want to know if something exists, see how it looks when you're in a good brain state. And if you want to know if a brain state is good, evaluate it according to how well it grows out of the life leading up to it, and how much it fosters joy and mattering. Neurath was right: We start with a conception of what matters and we don't give it up. But along the way, we meet other people and fall in love with them and have children, and we become their ancestors. As their ancestors, the web of genes and meaning that created us gives birth to them, and the things that sustain belief grow on the tree like apples.

Every belief, whether it be in quantum fields or Kris Kringle, is an event in a life. Life has a logic that is above and beyond the purely mechanical.* While a mechanism can break (assuming whether or not something is a mechanism can be abstracted

* See Michael Thompson's excellent *Life and Action: Elementary Structures of Practice and Practical Thought* for a discussion of how the logic of something alive differs from the logic of something inanimate. To give one example, "a

from the purpose of the living being who made it), living things pass from immaturity to maturity and reproduction and to old age, and finally, they die.

Death is something we're finally ready to talk about, because now that we've imagined the perfect eternal moment, we've also, by doing so, imagined the perfect last moment, because both moments—an eternal moment and the moment of death—have in common that they are the only remaining moment we are going to get. If we are able to face our last moment and give it a point or see its point, then we can give or see the point of our whole lives, because any moment could be our last.

So what should our attitude be to a moment that could be our last? Surprisingly, some insight comes from our old friend the neuroscientist Porges, by way of a horrible story he tells about an experiment some scientists performed on rats.

C. P. Richter in his article "On the Phenomenon of Sudden Death in Animals and Man" reports that he took some rats, "prestressed" them by cutting their whiskers, and threw them into a tank to swim until they drowned. A previous theorist, Walter B. Cannon, predicted that the rats would be in a state of fight-or-flight, sympathetic arousal until death, that they would die from the exhaustion of nonstop fear. And that turned out to be the case for tame rats, who swam in terror for several hours

---

sperm's tail is to help it fertilize the egg" is true, even if the overwhelming majority of sperm cells never fertilize an egg.

before dying. However, wild rats went into vagus shutdown and died in fifteen minutes, peacefully.

Now, which would you rather be—the wild rat or the tame rat? I'd rather be the wild rat. If I'm really at the end, I would rather go peacefully, mystically welcoming whatever the transition takes me to, even if it turns out to take me nowhere. It seems particularly awful that the tame rats were in a state of desperate terror and striving to the last moment. If we can respond to our death with peaceful acceptance, using a parasympathetic response rather than a sympathetic response, then we are responding to it with something like hope. Not a hope with a specific cognitive content—a specific belief that we will be in a green field, let's say—but a hope all the same. I like to think that a few minutes before my death, I will be in the mood for comedy, cracking jokes with friends and family members. But at the actual moment, I'd like to die as a mystic in vagal shutdown—free of thoughts, in love with the unknown.

CONCLUSION

My Future Grandson

I don't know what happens next, but I can imagine, and in a few different ways. If I use my scientific head, I can imagine some future evolution of humanity that includes robotic intelligence and heightened versions of our favorite animals and plants—bears and dogs and horses and oak trees—choosing to reconstitute me and tell me what it was all about from their point of view. This book is my audition for that resurrection, but I promise I'll ask them to bring you along too. If I use my right brain a little more, I can believe that there was something like love at the bottom of it all, and something like love in its fulfillment. I don't know, but presents need to be wrapped for them to be unwrapped, or they wouldn't be presents, so it's just as well I don't.

More concretely, I can imagine going to see Tammi some

years from now, telling her I'm sorry my kid made her kid stop believing in Santa Claus, because now I kind of do. I think Santa Claus is a manifestation of the Limitless as it presents the Gentle Face. And I let her read the foregoing and this is what she says:

"To be honest, Eric, I have zero interest in logic or mysticism or philosophy. I like comedy, but I don't take it seriously—it's just for fun, right? I don't know anything about the Ari or his Gentle Face—they both sound weird and alien. I'm from Nashville and, next to my family, I'm most interested in my world-class collection of cute figurines of mice playing musical instruments. I don't think everybody should have a collection, but I am proud of mine. When I play with them, they are precious. I don't want to take them out into the world where they might get lost or broken.

"I have a figurine of a Santa Claus mouse. His name is 'Santa Whisker Paws.' I love him and I like to play with him, but I am always careful not to break him.

"When I told my kids about Santa Claus, I was letting them in on something I think is precious and that I don't want to

bring out into the world, just like I don't want to bring my mouse figurines into the world. I won't show my mouse figurines to everybody because I don't want to be made fun of. But in my home, I love them. And in fact, Santa Claus brought my daughter Santa Whisker Paws."

Why am I discussing this? Imagine Tammi almost gave up on me because of the whole incident with Ari, Schuyler, and Santa Claus at the zoo, but then, for whatever reason, she didn't, and Ari went from being friends with her son, Schuyler, to being friends with her daughter, Kendra. They started going out, and after adolescent intimacies and betrayals, their relationship developed into love and marriage.

So now Tammi is the grandma and I am the grandpa of a child. And someday this child, grown old himself, will try to make sense of his upbringing. And he will look back and see a grandpa who believes in the Gentle Face of the Limitless, and a grandma who believes in Santa Claus, and a dawn in which ideas, now clear to him, first appeared through the mist of Christmas morning as presents in a sack, labeled LOGIC and MYSTICISM and COMEDY. And as he unwraps them, he will say, and imagine me saying with him, "Ho Ho Ho."

IN THE END
BEGINNING
IS
WEDGED IN
THE
IS
WED

SUGGESTIONS FOR FURTHER READING

Ainsworth, Mary D. S., Mary C. Blehar, Everett Waters, and Sally Wall. *Patterns of Attachment: A Psychological Study of the Strange Situation.* New York: Psychology Press, 1979. A rigorous, empirical, therapeutically oriented development of attachment theory, which is an indispensable tool for understanding our emotional lives and, I believe, our religious lives.

Carroll, Lewis. "What the Tortoise Said to Achilles." *Mind* 14 (1895): 278–80. Kindle edition. A profound essay on the philosophy of logic, by the author of *Alice's Adventures in Wonderland.*

Chesterton, Gilbert K. *Orthodoxy: The Classic Account of a Remarkable Christian Experience.* Colorado Springs: Shaw Books, 2001. Chesterton was guru to J. R. R. Tolkien and C. S. Lewis and is a foundational source of both modern fantasy writing and Christian apologetics. This book is probably the best argument for Christianity I've ever read.

Conze, Edward, trans. *Buddhist Wisdom Books: The Diamond Sutra and the Heart Sutra.* New York: Harper & Row, 1972. Super-deep stuff on form, emptiness, and the emptiness of emptiness, and a foundational text for Mahayana Buddhism.

Davidson, Donald. *Inquiries into Truth and Interpretation.* New York: Oxford University Press, 1984. A master of American analytic philosophy and one of my advisors, Davidson takes on deep issues about language, truth, and the mind. I don't think he's right, but he is worth wrestling with.

Descartes, René. *Discourse on Method and Meditations on First Philosophy.* 4th ed. Trans. Donald A. Cress. Indianapolis: Hackett Publishing, 1999. I take a lot of swipes at Descartes in this book. You should read him if you haven't yet and see if you think I'm being unfair.

Dionysius the Areopagite. *The Mystical Theology and the Divine Names.* Trans. C. E. Rolt. Mineola: Courier Dover Publications, 2012. A foundational text for Western mysticism and negative theology—the view that you can't say what God is, but you can say what He's not.

Dreyfus, Hubert L. *Being-in-the-World: A Commentary on Heidegger's Being and Time, Division I.* Cambridge: MIT Press, 1991.

———. *What Computers Still Can't Do: A Critique of Artificial Reasoning.* Cambridge: MIT Press, 1992. My advisor at Berkeley, Dreyfus demolishes the idea that a computer with an explicit program could ever think, and by doing so, he puts forward an alternative account of what human intelligence is and, along the way, explains the anti-Cartesian phenomenologists Heidegger and Merleau-Ponty. *Being-in-the World,* his commentary on Division 1 of Heidegger's *Being and Time,* is also highly illuminating.

Eliade, Mircea. *The Myth of the Eternal Return: Cosmos and History.* Trans. Willard R. Trask. Princeton: Princeton University Press, 2005. Like Chesterton, Eliade is both a major fantasy writer and an important theorist of religion. His phenomenology of archaic religion gives us a sense of what living in a mythic society would be like, and what we lose when we become modern.

Elster, Jon. *Ulysses and the Sirens: Studies in Rationality and Irrationality*. New York: Cambridge University Press, 1979. Elster wrestles with issues of when and how it's rational to be irrational: intrinsic side benefits, binding of the will, and self-gaming.

Farrow, George W., and I. Menon. *The Concealed Essence of the Hevajra Tantra: With the Commentary Yogaratnamala*. Delhi: Motilal Banarsidass Publishers, 1992.The Hevajra Tantra is one of the highest yoga Tantras, the most esoteric texts of Vajrayana Buddhism. It lays out a deep, exciting, and sexually explicit analysis of how to get ultimate reality and regular reality working together.

Heidegger, Martin. *Being and Time*. Trans. John McQuarrie and Edward Robinson. New York: Harper & Row, 1962. A devastating criticism of Descartes and the whole Western philosophical tradition going back to Plato. Heidegger discovers that there are three different ways that something can be: Dasein (beings whose being is an issue for them—i.e., us), equipment (stuff like hammers), and entities (things we can consider in a context-independent way). Very important if you are worried about whether Santa Claus exists and, because of the astonishingly turgid writing, best explored with Hubert Dreyfus's *Commentary* under one elbow.

Hobbes, Thomas. *Leviathan*. Ed. Marshall Missner. New York: Pearson Education, 2008. A pessimistic, materialistic, authoritarian account of human nature and politics that, along the way, gives a pessimistic, materialistic, authoritarian account of the nature of comedy.

James, William. *The Varieties of Religious Experience: A Study in Human Nature*. London: Longmans, Green, 1902. A pragmatic approach to the study of religion by the founder of experimental psychology, who coined the phrase "stream of consciousness."

Kaplan, Aryeh. *Inner Space: Introduction to Kabbalah, Meditation and Prophecy*. 2nd ed. Ed. Abraham Sutton. New York: Moznaim Publishing, 1991. The best commentary I know on the teachings of the Ari in English.

———. "Paradoxes." *The Aryeh Kaplan Reader: Collected Essays.* Brooklyn: Mesorah Publications, 1983. 131–39.

Krishnamurti, J. *Freedom from the Known.* Ed. Mary Lutyens. New Delhi: B. I. Publications, 1959. A twentieth-century mystic who seems to have almost no ontological commitments.

Luzzatto, Moshe Chaim. *138 Openings of Wisdom.* Trans. Avraham Yehoshua Greenbaum. Jerusalem: Azamra Institute, 2005. Born in Italy, Rabbi Moshe Chaim Luzzato was a playwright, philosopher, and kabbalist in the 1700s. Luzzato's writings were so controversial that the rabbis of the day threatened him with excommunication and burned most of his writings. This one, a commentary on the Ari's Tree of Life, survives.

Magee, Glenn Alexander. *Hegel and the Hermetic Tradition.* Ithaca: Cornell University Press, 2008. Excellent bridge between kabbalah and Western philosophy by way of Hegel, who branches backward toward Kant and forward toward continental philosophy: Marx, Heidegger, Derrida, and friends.

McGilchrist, Iain. *The Master and His Emissary: The Divided Brain and the Making of the Western World.* New Haven: Yale University Press, 2012. Brilliant far-reaching account of the relationship between the two hemispheres throughout history.

Merleau-Ponty, Maurice. *Phenomenology of Perception.* Trans. Donald A. Landes. New York: Routledge, 2012. Amazing non-Cartesian account of the role of the body and perception in knowledge and life in general. It's important to read Merleau-Ponty to avoid getting hornswoggled by information-processing models of the mind.

Moore, Alan. *Promethea, Book 1.* New York: DC Comics, 2001. This graphic novel by the author of *Watchmen* engages seriously with the kabbalah and the question of whether and to what extent imagined realities are real.

Moore, Clement Clarke. *'Twas the Night before Christmas.* Riverside, NJ: Andrews McMeel Publishing, 2009. Important ur-text for Santa.

Nanamoli, Bhikkhu, and Bhikkhu Bodhi, trans. *The Middle Length Discourses of the Buddha: A Translation of the Majjhima Nikaya.* Somerville: Wisdom Publications, 1995. In these discourses, the ancient Indian thinker Siddhartha Gautama explains why life is suffering and what to do about it.

Nissenbaum, Stephen. *The Battle for Christmas.* New York: Vintage Books, 1997. A fascinating social history of Christmas, Santa Claus, and the connection between New York real estate and "A Visit from St. Nicholas."

Panksepp, Jaak, and Lucy Biven. *The Archaeology of Mind: Neuroevolutionary Origins of Human Emotions.* New York: W. W. Norton, 2012. Philosophically important work on the basis of the emotions in the brain and their importance for thought and action. Includes a jaw-dropping discussion of the nature of play and laughter in rats.

Peirce, Charles S. *Charles S. Peirce: Selected Writings.* Ed. Philip P. Wiener. Mineola: Dover Publications, 1966. Great American philosopher and teacher of William James, Peirce profoundly engages with the issue of what truth means in a non-Cartesian framework.

Porges, Stephen W. *The Polyvagal Theory: Neurophysiological Foundations of Emotions, Attachment, Communication, and Self-Regulation.* New York: W. W. Norton, 2011. A great book for self-understanding. Porges makes surprising connections between neuroscience and a wide range of theoretical, practical, and clinical issues.

Proust, Marcel. *In Search of Lost Time, Vol. VI: Time Regained.* Trans. Andreas Mayor and Terence Kilmartin. New York: Random House, 2000. Last volume of the author's exciting *In Search of Lost Time.* The title is a bit of a spoiler.

Quine, Willard Van Orman. "On What There Is." *From a Logical Point of View.* Cambridge: Harvard University Press, 1953. 1–19. Quine is important if you want to understand how analytic philosophy deals with questions of truth, language, and reality. Whether you want to understand that or not depends on you and how much else you have going on.

Roebuck, Valerie J., trans. *The Upanishads.* London: Penguin Books, 2003. Important ancient Indian dialogues on the relationship between your life and ultimate reality. It used to be that you would have to make a pilgrimage to a guru's hermitage and work for years to prove you're worthy to hear these teachings; now you can get them in a bookstore.

Russell, Bertrand. "On Denoting." *Mind* 14.56 (1905): 479–93. Seminal essay on how to use logic to sort out problems about what exists: For example, how can the sentence "The current king of France is bald" be true if there is no current king of France for it to refer to?

Scholem, Gershom. *Major Trends in Jewish Mysticism*. New York: Schocken Books, 1995. Still the best survey of kabbalah.

Siegel, Daniel J. *The Developing Mind*. New York: Guilford Press, 2012. Siegel is a psychiatrist who brings together mindfulness meditation, neuroscience, and attachment theory. I became acquainted with several of the neuroscientists discussed in this book by way of a conference he organized at UCLA.

Smullyan, Raymond M. *The Tao Is Silent*. New York: HarperCollins, 1977. Smullyan is both a logician and a profound philosopher of mysticism. The title of my first chapter on logic is an "homage" to his work. He's also funny, so if you're interested in logic, mysticism, and comedy, you would enjoy him.

Spence, Joseph. *Observations, Anecdotes, and Characters, of Books and Men*. 1820. Vol. 1. Ed. James M. Osborn. London, Walter Scott, 1966. Sect. 1259. 462. Quoted in David Brewster, *Memoirs of the Life, Writings, and Discoveries of Sir Isaac Newton*. Vol. 2. 1855. 407.

Tarski, Alfred. "The Concept of Truth in Formalized Languages." *Logic, Semantics, Metamathematics: Papers from 1923 to 1938*. Trans. J. H. Woodger. Indianapolis: Hackett Publishing, 1983. 152–278.

Taylor, Jill Bolte. *My Stroke of Insight: A Brain Scientist's Personal Journey*. New York: Viking Penguin, 2008. Amazing first-person account by a neuroscientist whose massive stroke made her feel one with everything. Depending on your point of view, this either confirms the truth of mysticism or completely debunks it.

Vimalakirti. *The Holy Teaching of Vimalakirti: A Mahayana Scripture*. Trans. Robert A. F. Thurman. University Park: Penn State University Press, 1976. A concise, imaginative, and readable Mahayana sutra. The translator, Robert Thurman, makes connections with Tantra, the Avatamsaka sutra, and the Prajnaparamita sutra.

Wasserstrom, Steven M. *Religion after Religion: Gershom Scholem, Mircea Eliade, and Henry Corbin at Eranos*. Princeton: Princeton University Press, 1999. A social history of the academic study of religion, Wasserstrom illuminates the spiritual perspective that informed the work of three scholars engaged with the question of the relevance of religion and myth in a post-Enlightenment culture.

Weber, Max. *The Protestant Ethic and the Spirit of Capitalism*. New York: Routledge, 2001. If you buy things, have a job, or have a bank account, Weber can help explain to you what's going on.

Wittgenstein, Ludwig. *Tractatus Logico-Philosophicus*. New York: Routledge, 2013. A great book about the distinction between what can be thought and what can't be thought. The author's sequel, *Philosophical Investigations*, is also good.

Wolfe, Gene. "And When They Appear." *Strange Travelers: New Selected Stories*. New York: Tom Doherty Associates, 2001.The most heartbreaking, beautiful story about whether Santa exists that I know.

Wordsworth, William. *Selected Poetry*. Ed. Mark Van Doren. New York: Random House, 1950. The philosopher John Stuart Mill had a major depressive episode at age twenty, brought on by thinking too much. He credits the restoration of his sanity to a reading of Wordsworth's poetry.

ACKNOWLEDGMENTS

This book is a product of a long, unplanned, and unconscious process of growth, so the list of people to thank for the book has considerable overlap with the list of people to thank for myself.

My parents—Benjamin and Charlotte Kaplan—and grandparents, Wolf and Giselle Buchsbaum and Edward and Dora Kaplan.

My wife, Raduca Kaplan, and children, Ari and Mira.

My guides in the world of mysticism: Phra Chalor Kosadhammo, Robert Thurman.

My guides in the world of philosophy: Mark Ast, Stanley Cavell, Sidney Morgenbesser, Bernard Williams, John Searle, Hubert Dreyfus.

My guides in the world of kabbalah and Judaism: David Sacks, R. Simcha Weinberg, R. David Friedman, R. Perets Auerbach, R. Avraham Sutton.

My editor, Stephen Morrow, and all the people at Dutton.

For reading early versions of this and offering invaluable comments—Charlie Buckholtz, Rabbi Simcha Weinberg, Noam Cohen.

My colleagues on *The Big Bang Theory,* especially its creators—Chuck Lorre and Bill Prady—show runner Steve Molaro; the writers, Dave Goetsch, Jim Reynolds, Steve Holland, Maria Ferrari, Tara Hernandez, and Jeremy Howe; and the cast: Mayim Bialik, Kaley Cuoco-Sweeting, Johnny Galecki, Simon Helberg, Kunal Nayyar, Jim Parsons, and Melissa Rauch.

INDEX

INDEX

INDEX

INDEX

INDEX